A MAPLE® Manual for

ENGINEERING MECHANICS

STATICS
COMPUTATIONAL EDITION

Robert W. Soutas-Little
Michigan State University

Daniel J. Inman
Virginia Polytechnic Institute and State University

Daniel S. Balint
Imperial College London

THOMSON
™

Australia · Canada · Mexico · Singapore · Spain · United Kingdom · United States

A Maple® Manual for Engineering Mechanics: Statics, Computational Edition
by Robert W. Soutas-Little, Daniel J. Inman, and Daniel S. Balint

Associate Vice President and Editorial Director:
Evelyn Veitch

Publisher:
Chris Carson

Developmental Editor:
Hilda Gowans

Permissions Coordinator:
Vicki Gould

Production Services:
RPK Editorial Services, Inc.

COPYRIGHT © 2008 by Nelson, a division of Thomson Canada Limited.

Printed and bound in the United States
1 2 3 4 07

For more information contact Nelson, 1120 Birchmount Road, Toronto, Ontario, Canada, M1K 5G4. Or you can visit our Internet site at
http://www.nelson.com

Library Congress Control Number:
2007920216

ISBN-10: 0-495-29606-6
ISBN-13: 978-0-495-29606-5

Copy Editor:
Pat Daly

Proofreader:
Erin Wagner

Indexer:
Shelly Gerger-Knechtl

Production Manager:
Renate McCloy

Creative Director:
Angela Cluer

ALL RIGHTS RESERVED. No part of this work covered by the copyright herein may be reproduced, transcribed, or used in any form or by any means—graphic, electronic, or mechanical, including photocopying, recording, taping, Web distribution, or information storage and retrieval systems—without the written permission of the publisher.

For permission to use material from this text or product, submit a request online at
www.thomsonrights.com

Every effort has been made to trace ownership of all copyright material and to secure permission from copyright holders. In the event of any question arising as to the use of any material, we will be pleased to make the necessary corrections in future printings.

Maple® is a registered trademark of Maplesoft, Waterloo, Ontario, Canada.

Interior Design:
RPK Editorial Services

Cover Design:
Andrew Adams

Compositor:
Integra

Printer:
Thomson/West

Cover Image Credit:
Rose Kernan

North America
Nelson
1120 Birchmount Road
Toronto, Ontario M1K 5G4
Canada

Asia
Thomson Learning
5 Shenton Way #01-01
UIC Building
Singapore 068808

Australia/New Zealand
Thomson Learning
102 Dodds Street
Southbank, Victoria
Australia 3006

Europe/Middle East/Africa
Thomson Learning
High Holborn House
50/51 Bedford Row
London WC1R 4LR
United Kingdom

Latin America
Thomson Learning
Seneca, 53
Colonia Polanco
11560 Mexico D.F.
Mexico

Spain
Paraninfo
Calle/Magallanes, 25
28015 Madrid, Spain

Contents

Introduction	**vii**
1 Using *Maple* Computational Software	**1**
Numerical Calculation	2
Working with Functions	2
Symbolic Calculations	5
Solving Algebraic Equations	6
Graphs and Plots	7
Application of *Maple* to a Statics Problem	9
Computational Window 1.1	10
2 Vector Analysis	**13**
Computational Solution—Sample Problem 2.5	13
Maple as a Vector Calculator	15
Computational Window 2.1—Vector Calculator	15
Computational Window 2.2—Vector Algebra	16
Computational Window 2.3—Creating a Unit Vector	17
Computational Window 2.4—Symbolic Processor	18
Computational Solution—Sample Problem 2.7	18
Computational Solution—Sample Problem 2.8	19
Solution of Simultaneous Linear Equations	20
Computational Window 2.5	20
Computational Solution—Sample Problem 2.9	21
Computational Solution—Sample Problem 2.10	21
Using *Maple* for Other Matrix Calculations	22
Computational Solution—Sample Problem 2.12	22
Scalar or Dot Product	23
Computational Window 2.6	23
Computational Solution—Sample Problem 2.13	24
Computational Solution—Sample Problem 2.14	25
Vector or Cross Product Between Two Vectors	26
Computational Window 2.7	26
Computational Window 2.8	27
Computational Solution—Sample Problem 2.16	27
Computational Solution—Sample Problem 2.17	28
Computational Solution—Sample Problem 2.18	29
3 Particle Equilibrium	**31**
Parametric Solutions	31
Computational Solution—Sample Problem 3.4	31
Computational Solution—Sample Problem 3.5	33
Computational Solution—Sample Problem 3.6	33
Computational Solution—Sample Problem 3.8	34
Solution of Nonlinear Algebraic Equations	34

Computational Solution—Sample Problem 3.9   35
Computational Solution—Sample Problem 3.10   36
Computational Solution—Sample Problem 3.13   36
Computational Solution—Sample Problem 3.13—Soft Springs   37
Computational Solution—Sample Problem 3.14   38
Computational Solution—Sample Problem 3.14—Vertical Deformation Only   39
Computational Solution—Sample Problem 3.15   40
Computational Solution—Sample Problem 3.15—Design Variation   42

4 Rigid Bodies: Equivalent Force Systems   44
Computational Solution—Sample Problem 4.4   44
Computational Solution—Sample Problem 4.5   44
Computational Solution—Sample Problem 4.6   45
Computational Solution—Sample Problem 4.7   45
Computational Solution—Sample Problem 4.8   45
Computational Solution—Sample Problem 4.9   46
Computational Solution—Sample Problem 4.10   46
Computational Solution—Sample Problem 4.11   46
Computational Solution—Sample Problem 4.12   47
Computational Solution—Sample Problem 4.13   47
Computational Solution—Sample Problem 4.14   48
Computational Solution—Sample Problem 4.16   48
Computational Solution—Sample Problem 4.17   49

5 Distributed Forces: Centroids and Center of Gravity   50
Numerical and Symbolic Integration   50
Computational Window 5.1   50
Computational Solution—Sample Problem 5.1   50
Computational Solution—Sample Problem 5.3   51
Three-Dimensional Scatter Plots   53
Computational Window 5.2   53
Computational Solution—Sample Problem 5.4   54

6 Equilibrium of Rigid Bodies   56
Computational Solution—Sample Problem 6.4   56
Computational Solution—Sample Problem 6.5   58
Computational Solution—Sample Problem 6.6   59
Computational Solution—Sample Problem 6.6—Symbolic Generation
   of Equilibrium Equations   60
Computational Solution—Sample Problem 6.6—Solution of the Six Equations
   of Equilibrium   61
Computational Solution—Sample Problem 6.7—Symbolic Generation and Solution
   of Equilibrium Equations   62
Computational Solution—Sample Problem 6.8—Computational Generation
   of Equilibrium Equations   65
Computational Solution—Sample Problem 6.8   67

Computational Solution—Sample Problem 6.10 68
Computational Solution—Sample Problem 6.11 69
Computational Solution—Sample Problem 6.12—Symbolic Generation
  of Equilibrium Equations 69
Computational Solution—Sample Problem 6.12—Numerical Solution 71
Computational Solution—Sample Problem 6.12—Numerical Solution
  for the Tension in the Cable 72

7 Analysis of Structures 74
Computational Window 7.1 74
Computational Window 7.2 75
Computational Window 7.3 77
Computational Solution—Figure 7.21—Method of Sections 78
Computational Solution—Sample Problem 7.2—Case 1 78
Computational Solution—Sample Problem 7.2—Case 2 79
Computational Solution—Sample Problem 7.3—Motor Torque Graph (L_2 = 200 mm) 80
Computational Solution—Sample Problem 7.3—Motor Torque Graph (L_2 = 300 mm) 81
Computational Solution—Sample Problem 7.4—Equations (a) and (b) 82
Computational Solution—Sample Problem 7.4—Equations (a) and (c) 82
Computational Solution—Sample Problem 7.4—Equations (b) and (c) 82

8 Internal Forces in Structural Members 84
Computational Solution—Sample Problem 8.4—Symbolic Solution 84
Computational Solution—Sample Problem 8.4—Numerical Solution
  (Linear Loading) 84
Computational Solution—Sample Problem 8.4—Numerical Solution
  (Sinusoidal Loading) 86
Computational Solution—Sample Problem 8.6—Maximum Moment 88
Computational Solution—Sample Problem 8.6—Shear and Moment Diagrams 90
Discontinuity Functions 92
Computational Solution—Sample Problem 8.7—Shear and Moment Diagrams 93
Computational Solution—Sample Problem 8.7—Alternate Method 94
Cables 96
Computational Solution—Sample Problem 8.8—L = 33 ft (Cable Length) 96
Computational Solution—Sample Problem 8.8—L = 31 ft (Cable Length) 97
Computational Solution—Sample Problem 8.9 97
Computational Solution—Sample Problem 8.10 98

9 Friction 101
Computational Solution—Sample Problem 9.2 101
Computational Solution—Sample Problem 9.2—Solution as a Function
  of the Coefficient of Friction and the Inclination Angle of the Plane 102
Computational Solution—Sample Problem 9.3 103
Computational Solution—Sample Problem 9.4 104
Wedges 105
Computational Solution—Figure 9.10—Symbolic Evaluation of Wedge System 105

Computational Solution—Figure 9.10—Dependency of the Applied Force P on the Wedge Angle θ — 106
Computational Solution—Figure 9.10—Dependency of the Applied Force P on the Coefficient of Friction μ — 107
Belt Friction — 109
Ratio of Tensions versus the Coefficient of Friction — 109
Ratio of Tensions versus the Angle of Contact — 110
Ratio of Tensions versus Coefficient of Friction and Contact Angle — 110

10 Moments of Inertia — 112
Computational Window 10.1 — 112
Principal Second Moments of Area — 112
Computational Window 10.2—Second Moment of Area — 112
Computational Solution—Sample Problem 10.5 — 113
Eigenvalue Problem — 114
Computational Solution—Sample Problem 10.7—Eigenvalue Solution — 114

11 Virtual Work — 116
Computational Solution—Sample Problem 11.3 — 116
Computational Solution—Sample Problem 11.4—Slide Problem — 118
Computational Solution—Sample Problem 11.4—Slide Problem (Variation) — 120
Computational Solution—Sample Problem 11.7 — 122

Summary — 123

Index — 125

Introduction

This supplement provides instruction for solving Statics problems using *Maple* computational software. The student must note that before any attempt is made to use this software to solve a Statics problem, the problem must be correctly modeled. A free-body diagram must be constructed and the correct equations of equilibrium formulated. Computational software only reduces the numerical burden and makes it possible to conceptualize the solution through plots, and facilitates parametric studies which enable design and physical principals to be examined. *Maple* is a product of Maplesoft, 615 Kumpf Drive, Waterloo, Ontario, Canada, N2V 1K8. Many of the sample problems from the *Statics* text are solved in detail using *Maple* and are organized by chapter. If the student is familiar with the basic operations of *Maple*, the preliminary material may be skipped and the sample problems from the book may be examined immediately.

This supplement is intended to teach the reader how to solve Statics problems using *Maple*. You are encouraged to run your version of *Maple* and try the various steps described in this supplement as you read it. It is suggested that you work straight through the first chapter of the supplement using *Maple* to reproduce any calculations. This should familiarize you with the *Maple* environment and introduce the basic syntax. The supplement should then be referred to as needed when solving the homework problems and studying the text. This supplement is written in a style consistent with the *Maple* environment. While this supplement suggests ways to use *Maple* to enhance your understanding of Statics and teach you efficient computational skills, you should feel free to browse the *Maple* manual and create your own methods for solving Statics problems and for using *Maple*.

Quality technical documents can be created entirely within *Maple*. This manual was created in *Maple* and demonstrates this capability. As a consequence, the input and output formats presented in this manual are consistent with actual *Maple* input and output. Explanations are provided for the generation of symbols and operators that do not appear on the standard keyboard. Any input that is executed remains in memory and can be used for future calculations. Thus, the order in which input is executed is critical to successful calculations, but the order of the actual input is not (although it is desirable to order input in a logical fashion). Constants and functions are not recognized by *Maple* until the corresponding input has been executed.

This supplement consists of 11 chapters. The first chapter is a general introduction to *Maple* that concludes with a sample application and can be studied while reading the first chapter of the *Statics* text. This is followed by 10 more chapters, one for each of Chapters 2 through 11 of the text. In each of these chapters, appropriate *Maple* solutions are presented for the sample problems in the text.

1 Using *Maple* Computational Software

A general overview of *Maple* is given in this chapter including instructions for performing numerical calculations, defining and manipulating functions, performing symbolic calculations, solving algebraic equations, and plotting results. Text, graphics, and mathematical input and output are created in cells. Cells can contain mathematical input and output as well as formatted text which allows presentation quality reports to be created directly within *Maple*. Text can be created by selecting *Text* from the insert drop-down menu. Input can be continued across multiple lines using the keystrokes *[shift]+[enter]* and is executed using the keystroke *[enter]*. All input must be followed by a semicolon for it to be interpreted by *Maple* as valid mathematical input. Mathematical input can be created using the *Maple* syntax or standard math notation. Input using *Maple* syntax can be created by selecting *Maple Input* from the *Insert* drop-down menu. Input using standard math notation can be created by selecting *2-D Math* from the *Insert* drop-down menu. For example, one of the solutions of the general quadratic equation $ax^2 + bx + c = 0$ would be defined as *(-b+sqrt(b^2-4*a*c))/(2*a)* using *Maple* syntax but could be written as $-b + \sqrt{b^2 - 4 \cdot a \cdot c}/2a$ using standard math notation. Input using *Maple* syntax appears directly at the command prompt and is created by default. Output may be suppressed by ending a line of input with a colon instead of a semicolon.

Mathematical documents can be created in *Maple* in *Worksheet* mode or *Document* mode. This manual is created in *Worksheet* mode with *Maple* syntax as this is the traditional way of defining input in *Maple* and is therefore compatible with older versions of the software. However, the *Document* mode allows for more freedom in typesetting input and output. Sophisticated typesetting can be done in *Document* mode, or in *Worksheet* mode with *2-D Input*, using *Palettes*, which can be accessed from the *View* drop-down menu. For example, integrals can be evaluated using the integral symbol $\int$ rather than the integration command. Further details on typesetting can be found using the *Help* drop-down menu.

Two types of definitions can be made using *Maple*. Assignments are made using a colon followed by an equals sign (:=). This type of definition assigns *rhs* to *lhs*, which allows *lhs* to be used in calculations. Constants and functions are common examples that use this type of definition. An equals sign (=) is used as a binary operator for logical operations and for defining equations. Equations are defined using the general syntax *lhs = rhs*. An equation can be assigned to a variable name using both types of definitions and the general syntax *name:= lhs = rhs*. Calculations can then be performed using *name* to represent the equation.

Numerical Calculation

The following is a simple numerical calculation that can be performed using any calculator but can also be performed using *Maple*. One root of the general quadratic equation is evaluated for values of $a = 2$, $b = 4$, and $c = 1$ using *Maple*. *Maple* can be used as a simple calculator by entering the values into the quadratic formula as follows. Addition, subtraction, multiplication, and division are performed in *Maple* using the standard keys that correspond to these operations.

```
> (-4+sqrt(4^2-4*2))/2*2;
```
$$-4 + 2\sqrt{2}$$

The constants a, b, and c can also be predefined, allowing the quadratic formula to be entered symbolically. Note that the constants are defined using a colon followed by an equals sign (:=).

```
> a:=2:
> b:=4:
> c:=1:
> (-b+sqrt(b^2-4*a*c))/(2*a);
```
$$-1 + \frac{1}{2}\sqrt{2}$$

An alternate approach is to define the quadratic equation as a function and then find the roots of this function (values of the independent variable at which the function is equal to zero). *Maple* often requires an initial guess for the root or a restricted domain to search for the root in and uses the *fsolve* function to solve numerically for the root using an iterative method. The syntax used to define and refer to functions is discussed in the *Working with Functions* section of this manual. The *fsolve* function is discussed in detail in the *Solving Algebraic Equations* section.

```
> f:=x->2*x^2+4*x+1:
> fsolve(f(x),x=-1..0);
```
$$-.2928932188$$

Another solution to the quadratic equation exists. This solution can be determined by evaluating $-b-\sqrt{b^2 - 4 \cdot a \cdot c}/2a$ but can also be obtained by choosing a different search domain for the root.

```
> fsolve(f(x),x=-5..-1);
```
$$-1.707106781$$

Working with Functions

Functions of one or more than one independent variable may be defined and manipulated using *Maple*. Functions are defined using the syntax $f:=(var1,var2,...) \rightarrow function$ but are referenced using the syntax $f(var1,var2)$. A function was defined and referenced using *Maple* syntax in the preceding quadratic equation example. Functions can be defined and evaluated at a finite number of points over a particular domain of the independent variable using the *array* function and a *for* loop. The array function creates one-dimensional and multi-dimensional arrays. The syntax *array(1..n)* is used to create a one-dimensional array where *n* is the number of elements in the array. The first element of a *Maple* one-dimensional array has an index of one. Elements of an array are referenced using the syntax *a[i]* where *a* is the name of the array and *i* is the index of the desired element. Operations can be repeated using a *for* loop and the general syntax *for i from 1 to n do; body; od;* where *n* is the number of times *body* (which can be any group of *Maple* commands) is executed. In the following example, two functions of a single independent variable *x* are defined and evaluated between −2 and 2 at increments of 0.1.

```
> f:=x->1+x^3:
> g:=x->sin(Pi*x):
```

A third function can be defined as the product of the two.

```
> w:=x->f(x)*g(x):
```

These functions can be evaluated at each point in the specified domain of *x* using the array function and a *for* loop. Note that the function *evalf* must be used to instruct *Maple* to return a floating point approximation instead of an exact result. The *evalf* function is discussed in greater detail throughout this manual.

```
> valf:=array(1..201):
> valg:=array(1..201):
> valw:=array(1..201):
> for i from 1 to 200 do
    valf[i]:=f((i-1)*.01);
    valg[i]:=g((i-1)*.01);
    valw[i]:=w((i-1)*.01);
  od:
> evalf(valf[7]);
```
$$1.000216$$

```
> evalf(valg[7]);
```
$$0.1873813146$$

```
> evalf(valw[7]);
```
$$0.1874217890$$

These functions can also be differentiated, integrated, and evaluated at points in the specified domain. Definite integration of a function *f(x)* is performed using the syntax *int(f(x),x=lowerlimit..upperlimit)* and indefinite integration is performed using the syntax *int(f(x),x)*. The derivative of a function *f(x)* can be determined using the syntax *diff(f(x),x)*. Higher order derivatives can be easily defined by nesting *diff* functions. For example, the second derivative of *f(x)* would be determined by *diff(diff(f(x),x),x)*. Note that the *subs* function must be used to evaluate the function *dw* at a specific *x* value. This is necessary because the function *w(x)* must be differentiated with respect to *x* before the specific value of *x* is substituted into the function *dw(x)*. Using the *subs* function preserves this order. The *subs* and *evalf* functions are used frequently throughout this manual.

```
> dw:=x->diff(w(x),x):
> plot([x,dw(x),x=0..2],labels=[`x`,`dw(x)`],color=black);
```

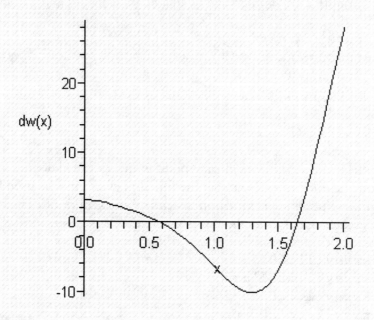

```
> evalf(subs(x=.9,dw(x)));
```
$$-4.415050521$$

```
> iw:=x->int(w(zeta),zeta=-2..x):
> plot([x,iw(x),x=0..2],labels=[`x`,`iw(x)`],color=black);
```

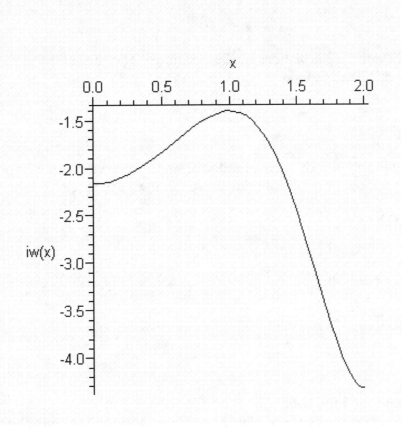

```
> evalf(subs(x=.9,iw(x)));
```
$$-1.426314754$$

Symbolic Calculations

Calculations can also be performed symbolically using *Maple*. The following example determines the derivative of the function *w(x)* symbolically. *Maple* performs calculations symbolically whenever possible by default. Thus no additional syntax is required to produce symbolic results. If numerical calculations are performed, an exact result is returned whenever possible by default. However, exact results are often extremely complex and difficult to interpret. For this reason, floating point approximations can be forced using the syntax *evalf(input)* where *evalf* is a function that returns a floating point approximation of *input*. Some numerical functions also exist such as the numerical equation solver *fsolve* and the numerical integration function *Int* that return only numerical results.

```
> diff(w(x),x);
```
$$3x^2 \sin(\pi x) + (1 + x^3) \cos(\pi x) \pi$$

```
> int(f(zeta),zeta=-2..x);
```
$$x - 2 + \frac{1}{4} x^4$$

Many other symbolic calculations can be performed. The cube of the sum of two variables is calculated as follows.

```
> (x+y)^3;
```
$$(x + y)^3$$

Since the input was already in simplest form, *Maple* returned it as the output. The function *expand* can be used to display the result in expanded form.

```
> expand(%);
```
$$x^3 + 3 x^2 y + 3 x y^2 + y^3$$

The function *factor* can be used to express the result in its simplest form. Note that the percent symbol (%) can be used to represent the last output.

```
> factor(%);
```
$$(x + y)^3$$

Solving Algebraic Equations

The *fsolve* function is a powerful *Maple* numerical function. The *fsolve* function numerically solves a system of one or more than one equation for the corresponding unknowns. Before the *fsolve* function is used, all known constants must be defined. Initial guesses for the unknowns or restricted domains in which to search for the unknowns are often required and are to be specified within the *fsolve* declaration (highly nonlinear equations require that the set of guesses be close to the solution or that the search domains contain only one solution and are sufficiently restricted). An initial guess or restricted search domain does not need to be specified if fsolve converges to the correct solution without one. However, it is always a good idea to supply a guess or a restricted search domain if possible.

Complicated equations and systems of equations require an initial guess or a restricted search domain for convergence to a solution to be achieved. The *fsolve* syntax is *fsolve({eq1,eq2,...},{var1=guess1,var2=guess2,...})* if initial guesses are provided and *fsolve({eq1,eq2,...},{var1,var2,...{var1=leftval1..rightval1,var2= leftval1..rightval2,...})* if search domains are specified. If the system of equations is large, it is helpful to assign each equation to a variable and use these variables in the *fsolve* declaration. Note that the search domains in the following example had to be chosen very close to the solutions for *fsolve* to return a solution. This, in general, is very difficult and illustrates the limitations that may come into play when solving systems of nonlinear algebraic equations.

```
> eq1:=3*x+4*cos(y)=1:
> eq2:=x-sin(y)=2:
> fsolve({eq1,eq2},{x=1..2,y=0..4});
```
$$\{x = 1.400000618, y = 3.785092990\}$$

The *solve* function is the symbolic counterpart of the *fsolve* function and can be used to solve equations and systems of equations symbolically using the same general syntax as *solve* except without initial guesses or search domains.

Graphs and Plots

Graphing calculators and computational software packages such as *Maple* can be used to graph functions and to facilitate the solution of difficult algebraic equations using graphs and plots. *Maple* can create two and three dimensional plots in a variety of coordinate systems including Cartesian, polar-cylindrical and spherical. Many other types of plots can be generated using *Maple* such as parametric plots and plots of implicitly defined functions. The most common type of plot encountered in Statics is a two-dimensional plot in a Cartesian coordinate system in which a function of one independent variable is plotted against the independent variable. The function *plot* can be used to generate this type of plot. The domain of interest must be specified and the plot can be formatted using options that follow the general syntax *option=value*. Many options exist for formatting plots, several of which are presented in this manual.

Maple uses an adaptive plotting algorithm that automatically picks a reasonable number of equally spaced points in the plotting interval at which the function is to be evaluated. Intervals over which the function has a high degree of fluctuation are further refined by *Maple*. Although suitable results are generally produced, it is possible that the plot will be too coarse. *Maple* supports the further refinement of plots through the option *numpoints* which defines the minimum total number of points to be generated. If *numpoints* is set high enough, it forces *Maple* to produce a more refined plot. Multiple plots can also be generated on the same graph. The general syntax for *plot* is *plot([[x,f1(x),x=leftval..rightval], [x,f2(x),x=leftval..rightval],...],option1=val1,...)*. The *plot* command is flexible and can be used with variations in the syntax and can be used differently to produce different types of plots. These capabilities are demonstrated later in this manual. An example of a two-dimensional plot in a Cartesian coordinate system is the following.

```
> plot([x,w(x),x=-2..2],x=-2..2,y=-5..5,
  title=`Two-Dimensional Cartesian Plot of
  w(x)`,labels=[`x`,`w(x)`],color=black);
```

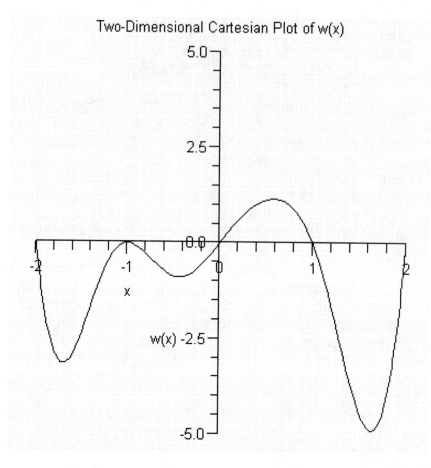

Polar plots can be created when the radius *r* is defined as a function of the angle θ (in radians) using the *plot* function with the option *coords* set equal to *polar*. The following is an example that demonstrates how to generate this type of plot.

```
> r:=theta->(theta/Pi)*sin(theta):
> plot([r(theta),theta,theta=-2*Pi..2*Pi],
  coords=polar,color=black);
```

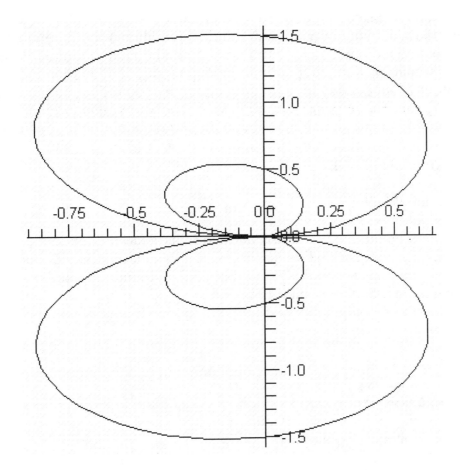

Note that the plot is constructed in a two-dimensional Cartesian coordinate system and that the horizontal and vertical axes represent the quantities $r\cos(\theta)$ and $r\sin(\theta)$, respectively.

Memory can be cleared in *Maple* using the syntax:

```
> restart;
```

This is particularly important when performing unrelated calculations that may use the same variable names within a single *Maple* session.

Application of *Maple* to a Statics Problem

An example of how to use *Maple* to solve a Statics problem is shown in Computational Window 1.1. The equation solved in this example is not difficult to solve and could have been easily solved by hand or using a calculator. The loading function is graphed on the beam to determine where the load is a maximum. To use *Maple* as a graphical calculator, define any constants in the function, define the function, and then plot the function over the desired domain of the independent variable. The maximum value of the function is determined by setting the derivative of the function equal to zero and solving for x.

The constant L is defined to be 10 ft and the function is defined according to the problem statement. The domain of x over which the function is evaluated is defined within the plot command and ranges between 0 and 10. The maximum load can be approximated graphically by plotting the derivative of the loading function $w(x)$ and noting the x value at which the derivative is zero and examining the value of the loading function at this x value. To determine more accurately the x value at which the load reaches a maximum value, the *fsolve* function must be used to find the root of the derivative of the load function. The maximum load is then determined by evaluating the loading function at the x value determined by *fsolve*.

Computational Window 1.1

This is a general example of how to use *Maple* to define a function, graph a function, differentiate a function, and find the maximum value of a function. Loading on a beam in lb/ft is given by the function $w(x) = 10 (x - x \sin[x/L])$. Determine where the load is a maximum if the length of the beam is 10 ft.

Define the length of the beam.
```
> L:=10:
```

Define the load function.
```
> w(x):=10*(x-x*sin(x/L)):
```

The load function can be graphed in the following manner.
```
> plot(w(x),x=0..10,y=0..40,labels=[`x`,`w(x)`],
  color=black);
```

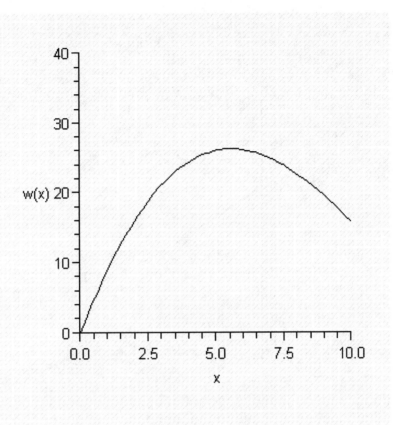

To determine where the load is a maximum, differentiate the function, set it equal to zero, and solve. An initial guess for the root must be supplied to the equation solver and can be easily obtained from the graph of the derivative of the load function.

```
> dw(x):=diff(w(x),x):
> plot(dw(x),x=0..10,y=-10..20,labels=[`x`,`dw(x)`],
  color=black);
```

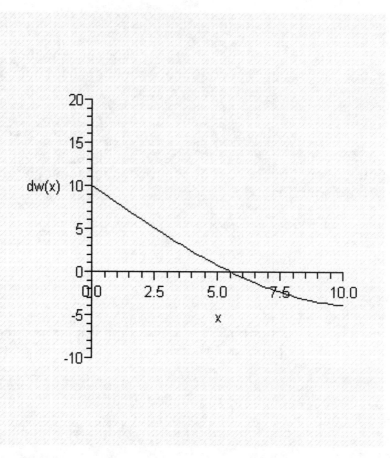

Find the root of the derivative of the load function, noting from the graph that the root is approximately $x = 6$.

> **xmax:=fsolve(dw(x)=0,x);**

$$xmax := 5.559684307$$

The maximum load is determined by evaluating the load function at the root of the derivative of the load function.

> **eval(subs(x=xmax,w(x)));**

$$26.25471277$$

The maximum load is 26.3 lb/ft.

2 Vector Analysis

In this chapter, detailed solutions of sample problems where *Maple* is useful are shown. These solutions show how *Maple* can be used as a vector calculator and show how systems of linear equations can be solved using matrix methods. Most of the problems in the second chapter of the text do not require the use of *Maple*, but you are encouraged to solve both the vector equations and the linear systems using *Maple* as this will give you more time to concentrate on the free-body diagrams and on forming the equations of equilibrium. You will also find that numerical errors are nearly eliminated when *Maple* is used to perform calculations. *Maple* can be used efficiently to solve Statics problems by observing the following procedure: (i) draw a free-body diagram for the particle and show all forces acting on the particle and show any specified geometry (ii) express each force vector in the given or desired coordinate system (iii) write the required vector equations (resultant or equilibrium) (iv) enter this information into a *Maple* worksheet and use *Maple* to perform all numerical calculations.

Sample Problem 2.5 requires the solution of a transcendental equation. This equation can be solved by hand as shown in the text by using trigonometric identities to obtain a quadratic equation for $\cos(\beta)$. Many algebraic equations of this nature cannot be easily solved and are best approached numerically. To solve an equation of this nature using *Maple*, first graph the function and determine approximations for the roots. The appropriate search domains for *fsolve* can be determined from the approximate roots. To gain familiarity with graphing and the *fsolve* and *evalf* functions, you should reproduce this solution on your own computer. Change some of the constants and examine the change in the graph and change in the root. These changes can be observed by altering the constant values in the solution and re-evaluating all input statements.

A general solution could have been obtained by defining $\theta = 70(\pi/180)$, $V = 500$, and $A = 320$. These values could then be easily altered without editing the transcendental equation. Consider Sample Problem 2.5 that required the solution of a transcendental equation. This equation can be solved by hand as shown in the text by using trigonometric identities to obtain a quadratic equation for $\cos(\beta)$. Many algebraic equations of this nature cannot be easily solved and are best approached numerically. To solve an equation of this nature, first graph the function and determine the approximate roots. The operator *fsolve* looks for all roots of a function. It is therefore necessary to specify which root is desired. This can be accomplished by restricting the domain in which zeros are sought. The appropriate domain restriction is determined by examining the graph.

Computational Solution — Sample Problem 2.5

> `restart;`

First define the function. For plotting and numerical calculations, built-in constants may need to be converted to floating point approximations through the use of the function *evalf*.

```
> f(beta):=500*sin(70*(evalf(Pi)/180))*cos(beta)-
  500*cos(70*(evalf(Pi)/180))*sin(beta)-
  320*sin(70*(evalf(Pi)/180)):
```

Plot the function to determine an approximation of the root.

```
> plot(f(beta),beta=0..100*(evalf(Pi/180)),y=-500..500,
  labels=[`beta (rad)`,`f(beta)`],color=black);
```

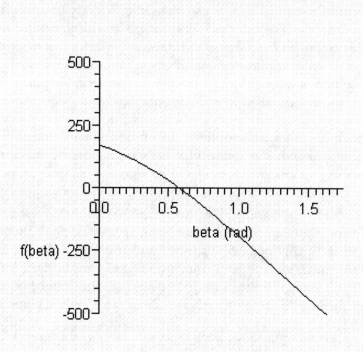

The root is shown to be approximately equal to 0.6 radians or 35 degrees. The numerical value of the root is determined in radians using the following *fsolve* command with the search domain restricted such that the correct root is found.

```
> SOLNrad=fsolve(f(beta),beta,0..1);
```
$$SOLNrad = 0.5764741152$$

The solution can be converted to degrees.

```
> SOLNdeg=(.5764741152)*(180/evalf(Pi));
```
$$SOLNdeg = 33.02953378$$

The exact root is 33 degrees and the force in newtons along b–b' is:

```
> Fb:=500*(sin(33*(evalf(Pi)/180)))/(sin(110*(evalf(Pi)/
  180)));
```
$$Fb := 289.7963776$$

Maple as a Vector Calculator

Force, moment, and position vectors are fundamental to the study of Statics. *Maple* can reduce the numerical labor of vector calculations but *Maple* should never be used to perform a vector operation that you do not fully understand. Remember that *Maple* is only a computational and graphing software program and cannot correct errors in the models or in the equations of equilibrium. Although the operations shown in Computational Windows 2.1 through 2.5 can be easily performed using a calculator and the scalar components of the vectors, these calculations are performed using *Maple* to demonstrate some of the powerful vector tools available in computational software packages.

Maple uses arrays for collecting numbers together. Vectors and matrices are essentially arrays and multidimensional arrays respectively that can be used with various vector and matrix functions. In *Maple*, elements of a vector begin with the index 1. The first element of a vector **a** in *Maple* is designated with the syntax *a[1]* and the first element of a matrix **A** is designated with the syntax *A[1,1]*. Vectors and matrices can be created using the *array* function and the general syntax *array(1..n)* and *array(1..n,1..m,...)* respectively where *1..n* is a range of indices. Vectors and matrices can also be defined directly using the syntax *[val1,val2,val3,...]* and *[[val11,val12...],[val21,val22...],...]* respectively.

Maple groups certain types of functions in what are referred to as packages. Packages generally contain groups of related functions that are not as commonly used as the standard set of *Maple* functions. Before these functions can be used, the correct package must be loaded using the syntax *with(packagename)*. The advanced vector and matrix functions are located in the *linalg* package. Two functions exist for defining vectors and matrices that are located in *linalg*. The function *vector* defines a vector using the syntax *vector(#,[val1,val2,...])* where # is the number of elements in the vector and *val1,val2,...* are the elements of the vector. The function *matrix* defines a matrix using the syntax *matrix(#r,#c,[val11,val12,...val21,val22,...])* where #r is the number of rows, #c is the number of columns, and *val11,val12,...* are the elements of the matrix. For matrices, elements are entered in the matrix function declaration across the rows from the upper left to the lower right. Vectors and matrices are defined in this manual using all three of the various methods and you should become familiar with all of them.

Computational Window 2.1 — Vector Calculator

The specific keystrokes or codes for performing vector calculations differ on the various software packages and you will need to select the specific software most suited for your calculations. However, all programs have certain common characteristics and these will be presented here. Vectors are treated either as column or row arrays of three numbers, with addition and subtraction performed component by component. An example of vector addition and subtraction is shown.

```
> restart;
> with(linalg):

> [10,-5,3]+[6,5,-2];
```
$$[16, 0, 1]$$

```
> [10,-5,3]-[6,5,-2];
```
$$[4, -10, 5]$$

The elements of vectors may contain mathematical expressions.

```
> evalf([5*sin(30*(evalf(Pi/180))),7/5,tan(50*
  (evalf(Pi/180)))]+[sqrt(2),5^3,21*6]);
```
$$[3.914213562, 126.4000000, 127.1917536]$$

The magnitude of a vector can also be computed using *Maple* commands. The magnitude of a vector is defined as the square root of the scalar product of the vector with itself. The *dotprod* function, which is located in the *linalg* package, is used to compute the scalar product in *Maple*. The scalar product (dot product) is discussed in detail in the *Scalar or Dot Product* section.

```
> evalf(sqrt(dotprod([10,-5,3],[10,-5,3])));
```
$$11.57583690$$

Computational Window 2.2 — Vector Algebra

When a vector is to be used in more than one operation, it is best to define a symbol for the vector and perform the vector operations using this symbol.

```
> restart;
> with(linalg):
> A:=[10,-5,3]: B:=[6,5,-2]:
> A+B;
```
$$[16, 0, 1]$$

```
> A-B;
```
$$[4, -10, 5]$$

```
> evalf(sqrt(dotprod(A,A)));
```
$$11.57583690$$

```
> evalf(sqrt(dotprod(B,B)));
```
$$8.062257748$$

In many cases, it is useful to define the result as another symbol and to write vector equations using this symbol.

```
> E:=A+B;
```
$$E := [16, 0, 1]$$

```
> F:=A-B;
```
$$F := [4, -10, 5]$$

The vector equations can contain more complex algebraic expressions.

```
> E:=2*A+B/5:
> evalf(E);
```
$$[21.20000000, -9., 5.600000000]$$

Computational Window 2.3 — Creating a Unit Vector

A unit vector in the direction of a vector can be easily obtained when symbols are used to represent vectors. It is obtained by dividing a vector by its magnitude.

```
> restart;
> with(linalg):
> A:=[5,2,-7]:
> magA:=evalf(sqrt(dotprod(A,A)));
```
$$magA := 8.831760866$$

```
> a:=(1/magA)*A;
```
$$a := [0.5661385170, 0.2264554068, -.7925939238]$$

A unit vector can also be obtained by using the *normalize* operator.

```
> evalf(normalize(A));
```
$$\begin{bmatrix} 0.5661385170 & 0.2264554068 & -.7925939238 \end{bmatrix}$$

Although these calculations are done using computational software, they may be done by hand and should be performed by hand the first time. The vector **A** can now be written as a magnitude multiplied by its unit vector.

```
> A:=(magA)*a;
```
$$A := [4.999999999, 2.000000000, -6.999999999]$$

The components of the unit vector are the direction cosines of the vector **A**.

Computational Window 2.4 — Symbolic Processor

Most computational software packages contain a symbolic processor that allows equations to be solved symbolically. Symbolic solution of an equation gives the general algebraic expressions that describe the solutions.

```
> restart;
> with(linalg):
```

All standard operations, such as the addition of two vectors, may be performed symbolically.

```
> [Ax,Ay,Az]+[Bx,By,Bz];
```
$$[Bx + Ax, By + Ay, Bz + Az]$$

The magnitude of the vector may be computed symbolically.

```
> A:=[Ax,Ay,Az]:
> (sqrt(dotprod(A,A)));
```
$$\sqrt{Ax\,Ax + Ay\,Ay + Az\,Az}$$

A vector may be multiplied by a scalar symbolically.

```
> c*[Ax,Ay,Az];
```
$$c\,([Ax, Ay, Az])$$

Computational Solution — Sample Problem 2.7

Sample Problem 2.7 asks that the resultant of the sum of two vectors and the unit vector in the direction of the resultant be determined. Sample Problem 2.8 involves a greater number of vector calculations and is an excellent example of the use of computational software.

```
> restart;
> with(linalg):
```

Define the vectors using algebraic symbols.

```
> A:=[6,2,0]:  B:=[0,7,5]:
> C:=A+B;
```
$$C := [6, 9, 5]$$

The magnitude of **C** can be computed directly.

```
> magC:=evalf(sqrt(dotprod(C,C)));
```
$$magC := 11.91637529$$

The unit vector in the direction of **C** is:

```
> c:=(1/magC)*C;
```
$$c := [0.5035088149, 0.7552632223, 0.4195906790]$$

Computational Solution — Sample Problem 2.8

```
> restart;
> with(linalg):
```

The position vectors from the origin to the point of attachment are (all units are in meters):

```
> A:=[-4,4,12]: B:=[-4,-6,12]: C:=[5,0,12]:
```

The magnitudes of **A**, **B**, and **C** are:

```
> magA:=evalf(sqrt(dotprod(A,A)));
> magB:=evalf(sqrt(dotprod(B,B)));
> magC:=evalf(sqrt(dotprod(C,C)));
```
$$magA := 13.26649916$$
$$magB := 14.$$
$$magC := 13.$$

The unit vectors along each of the cables are:

```
> a:=(1/magA)*A; b:=(1/magB)*B; c:=(1/magC)*C;
```
$$a := [-.3015113446, 0.3015113446, 0.9045340338]$$
$$b := [-.2857142857, -.4285714286, 0.8571428572]$$
$$c := [0.3846153846, 0., 0.9230769230]$$

The tensions in the cables are:

```
> TA:=368*a; TB:=259*b; TC:=482*c;
```
$$TA := [-110.9561748, 110.9561748, 332.8685244]$$
$$TB := [-74.00000000, -111.0000000, 222.0000000]$$
$$TC := [185.3846154, 0., 444.9230769]$$

The resultant of the three cable forces in newtons is:

```
> R:=TA+TB+TC;
```
$$R := [0.4284406, -0.0438252, 999.7916013]$$

Solution of Simultaneous Linear Equations
A system of linear equations is easily solved using matrix methods from linear algebra. Most problems in Statics result in such a system of linear equations and although they can be solved by hand, many calculators and all computational software packages will numerically or symbolically solve such systems of equations. Nonlinear equations also arise in mechanics and solution methods for such a system of equations will be discussed later in this supplement. Many hand calculators have the capability of handling matrices up to 6 × 6 in size or the ability to solve six equations with six unknowns. Computational software programs can solve much larger systems.

Computational Window 2.5

```
> restart;
> with(linalg):
```

Consider a system of three linear equations:

$2x + 3y - z = 6$
$3x - y + z = 8$
$-x + y - 2z = 0$

The coefficient matrix is:

```
> A:=matrix(3,3,[2,3,-1,3,-1,1,-1,1,-2]);
```

$$A := \begin{bmatrix} 2 & 3 & -1 \\ 3 & -1 & 1 \\ -1 & 1 & -2 \end{bmatrix}$$

The constant vector (which is the same as a 3 × 1 matrix) on the right side of the matrix equation is given by the following command.

```
> C:=matrix(3,1,[6,8,0]);
```

$$C := \begin{bmatrix} 6 \\ 8 \\ 0 \end{bmatrix}$$

The values of the unknowns x, y, and z are determined by left multiplying the constant vector by the inverse of the coefficient matrix.

```
> (x, y, z) = evalf(transpose(multiply(inverse(A), C)));
```
$$(x, y, z) = \begin{bmatrix} 3.066666667 & -.6666666667 & -1.866666667 \end{bmatrix}$$

Computational Solution — Sample Problem 2.9

```
> restart;
```

The solution of Sample Problem 2.9 involves three equations in three unknowns. Although computational software makes solving the problem easier, it can also be solved easily with only the help of a calculator.

```
> lambda_y:=evalf((500/300)*sin(30*Pi/180));
```
$$\lambda y := 0.8333333333$$

```
> lambda_x:=sqrt(1-lambda_y^2);
```
$$\lambda x := 0.5527707984$$

```
> lambda_xm:=-lambda_x;
```
$$\lambda xm := -.5527707984$$

```
> A:=evalf(500*cos(30*Pi/180)-300*lambda_x);
```
$$A := 267.1814625$$

```
> Am:=evalf(500*cos(30*Pi/180)-300*lambda_xm);
```
$$Am := 598.8439415$$

Maple has been used here simply as a numerical calculator.

Computational Solution — Sample Problem 2.10

```
> restart;
> with(linalg):
```

Sample Problem 2.10 is an example of a problem that yields a system of simultaneous equations. In the past, problems of this nature were solved by hand by eliminating all the unknowns but one in one equation and solving this equation for the unknown. This unknown is then substituted into the other equations and the unknowns are determined one by one.

Form vectors from the camera to the base of each leg.

```
> A:=[-24,-40,0]: B:=[21,-40,12]: C:=[21,-40,-12]:
```

Create unit vectors along each of the legs.

> `a:=evalf(normalize(A)); b:=evalf(normalize(B));`
> `c:=evalf(normalize(C));`

$$a := \begin{bmatrix} -.5144957555 & -.8574929256 & 0. \end{bmatrix}$$

$$b := \begin{bmatrix} 0.4492556773 & -.8557250999 & 0.2567175299 \end{bmatrix}$$

$$c := \begin{bmatrix} 0.4492556773 & -.8557250999 & -.2567175299 \end{bmatrix}$$

The coefficient matrix is formed using these three unit vectors.

> `Co:=matrix(3,3,[a[1],b[1],c[1],a[2],b[2],c[2],a[3],b[3],c[3]]);`

$$Co := \begin{bmatrix} -.5144957555 & 0.4492556773 & 0.4492556773 \\ -.8574929256 & -.8557250999 & -.8557250999 \\ 0. & 0.2567175299 & -.2567175299 \end{bmatrix}$$

The loading function is designated by the matrix **L**.

> `L:=matrix(3,1,[0,-50,0]);`

$$L := \begin{bmatrix} 0 \\ -50 \\ 0 \end{bmatrix}$$

> `[FA,FB,FC]=transpose(multiply(inverse(Co),L));`

$$([FA, FB, FC]) = \begin{bmatrix} 27.21110884 & 15.58132786 & 15.58132786 \end{bmatrix}$$

Using *Maple* for Other Matrix Calculations

Matrix addition, subtraction, and multiplication operations can be easily performed using *Maple*. Each matrix is defined by a symbol and then the matrix operations can be performed using these symbols. An example of this is shown in Sample Problem 2.12.

Computational Solution — Sample Problem 2.12

Although Sample Problem 2.12 can be solved by hand, all numerical operations can also be performed using *Maple*.

The *Maple* worksheet will look exactly like the solution that appears in the *Statics* text and the solution to the homework problems in this section can be generated completely using *Maple*. You should review the details of each matrix operation before using *Maple*.

```
> restart;
> with(linalg):
> A:=matrix(3,3,[2,-3,1,0,2,-1,3,1,1]);
```

$$A := \begin{bmatrix} 2 & -3 & 1 \\ 0 & 2 & -1 \\ 3 & 1 & 1 \end{bmatrix}$$

```
> B:=matrix(3,3,[5,3,2,-2,1,4,-1,0,-1]);
```

$$B := \begin{bmatrix} 5 & 3 & 2 \\ -2 & 1 & 4 \\ -1 & 0 & -1 \end{bmatrix}$$

```
> multiply(A,B);
```

$$\begin{bmatrix} 15 & 3 & -9 \\ -3 & 2 & 9 \\ 12 & 10 & 9 \end{bmatrix}$$

```
> multiply(B,A);
```

$$\begin{bmatrix} 16 & -7 & 4 \\ 8 & 12 & 1 \\ -5 & 2 & -2 \end{bmatrix}$$

Scalar or Dot Product

The scalar or dot product of two vectors can be calculated by use of the *dotprod* operator. This operation may be performed symbolically or numerically by pre-operating on the entire expression with the *evalf* operator. Applications of the dot product are discussed in the text and the basic operation is shown in Computational Window 2.6.

Computational Window 2.6

```
> restart;
> with(linalg):
```

Consider two vectors **A** and **B**.

```
> A:=[3,7,-1]: B:=[-5,4,9]:
```

The dot product of **A** and **B** is:

```
> dotprod(A,B);
```
$$4$$

The angle between the two vectors in radians is:

```
> magA:=evalf(sqrt(dotprod(A,A))):
> magB:=evalf(sqrt(dotprod(B,B))):
> theta:=evalf(arccos((dotprod(A,B))/(magA*magB)));
```
$$\theta := 1.523631842$$

The same angle in degrees is:

```
> theta*evalf(180/Pi);
```
$$87.29767406$$

The scalar or dot product can also be performed by defining the vectors as 3×1 matrices and performing matrix multiplication.

```
> A:=matrix(3,1,[3,7,-1]): B:=matrix(3,1,[-5,4,9]):
```

The dot product of **A** and **B** is (note that a scalar is the same as a 1×1 matrix):

```
> multiply(transpose(A),B);
```
$$\begin{bmatrix} 4 \end{bmatrix}$$

The vectors **A** and **B** may also be defined symbolically.

```
> A:=[Ax,Ay,Az]: B:=[Bx,By,Bz]:
```

The symbolic dot product is:

```
> dotprod(A,B);
```
$$Ax\,Bx + Ay\,By + Az\,Bz$$

Sample Problems 2.13 and 2.14 are solved using *Maple*. Each only requires the use of specific vector operations.

Computational Solution — Sample Problem 2.13

```
> restart;
> with(linalg):
```

We will first define the two vectors.

```
> A:=[3,-2,-2]: B:=[4,4,0]:
```

Scalar or Dot Product

The angle between **A** and **B** can be computed in radians as follows.

```
> magA:=evalf(sqrt(dotprod(A,A))):
> magB:=evalf(sqrt(multiply(B,B))):
> theta:=arccos((dotprod(A,B))/(magA*magB));
```
$$\theta := 1.398445737$$

The same angle in degrees is:

```
> theta*evalf(180/Pi);
```
$$80.12503859$$

The steps in the hand calculation can be checked if the results of intermediate calculations are displayed.

```
> dotprod(A,B);
```
$$4$$

```
> magA:=evalf(sqrt(dotprod(A,A)));
```
$$magA := 4.123105626$$

```
> magB:=evalf(sqrt(dotprod(B,B)));
```
$$magB := 5.656854248$$

```
> dotprod(A,B)/(magA*magB);
```
$$0.1714985852$$

Computational Solution — Sample Problem 2.14

Define the force vector and the relative position vector. Note that in vector solutions using *Maple*, relative vectors are written in the form **rBA** rather than $r_{B/A}$ as *Maple* does not allow a mathematical operator in the name of a variable.

```
> restart;
> with(linalg):
> F:=[10,-10,5]: rBA:=[12,3,4]:
```

A unit vector along the line is computed as follows.

```
> magrBA:=evalf(sqrt(dotprod(rBA,rBA))):
> n:=(1/magrBA)*rBA;
```
$$n := [0.9230769230, 0.2307692308, 0.3076923077]$$

The vector component of the force along the line is:

```
> FAB:=(dotprod(F,n))*n;
```
$$FAB := [7.810650886, 1.952662722, 2.603550295]$$

The vector component of the force perpendicular to the line is:

```
> Fp:=F-FAB;
```
$$Fp := [2.189349114, -11.95266272, 2.396449705]$$

Vector or Cross Product Between Two Vectors

The cross product between two vectors can be performed in *Maple* by using the operator *crossprod*. This will be demonstrated in Computational Windows 2.7 and 2.8.

Computational Window 2.7

Recall that the dot product could be obtained by a matrix multiplication between two vectors written as column or row matrices. The cross product cannot be obtained through any matrix multiplication. Consider three vectors **A**, **B**, and **C**.

```
> restart;
> with(linalg):
> A:=[3,-1,2]: B:=[-1,5,2]: C:=[2,1,2]:
```

Recall that the cross product is not commutative. Note that **A** × **B** is the negative of **B** × **A**.

```
> crossprod(A,B);
```
$$\begin{bmatrix} -12 & -8 & 14 \end{bmatrix}$$

```
> crossprod(B,A);
```
$$\begin{bmatrix} 12 & 8 & -14 \end{bmatrix}$$

The triple scalar (or box) product and the triple vector product are computed as follows. Note that the triple scalar product of three vectors is the same for all cyclic permutations of the three vectors (in other words, **A** · (**B** × **C**), **B** · (**C** × **A**), and **C** · (**A** × **B**) are equivalent).

```
> dotprod(A,(crossprod(B,C)));
```
$$-4$$

```
> dotprod(B,(crossprod(C,A)));
```
$$-4$$

```
> crossprod(A,(crossprod(B,C)));
```
$$\begin{bmatrix} -1 & 49 & 26 \end{bmatrix}$$

Scalar or Dot Product

The triple vector product can also be computed using a common identity.

```
> (dotprod(C,A))*B-(dotprod(A,B))*C;
```
$$[-1, 49, 26]$$

In Computational Window 2.8, the triple vector product $\mathbf{A} \times (\mathbf{B} \times \mathbf{C})$ and the triple scalar product $\mathbf{A} \cdot (\mathbf{B} \times \mathbf{C})$ are evaluated numerically.

Computational Window 2.8

The vector operations can also be performed symbolically.

```
> restart;
> with(linalg):
> A:=[Ax,Ay,Az]: B:=[Bx,By,Bz]: C:=[Cx,Cy,Cz]:
> crossprod(A,B);
```
$$\begin{bmatrix} Ay\,Bz - Az\,By & Az\,Bx - Ax\,Bz & Ax\,By - Ay\,Bx \end{bmatrix}$$

The triple scalar product is:

```
> dotprod(A,crossprod(B,C));
```
$$Ax\,By\,Cz - Bz\,Cy + Ay\,Bz\,Cx - Bx\,Cz + Az\,Bx\,Cy - By\,Cx$$

The triple vector product is:

```
> crossprod(A,crossprod(B,C));
```
$$\Big[[Ay\,(Bx\,Cy - By\,Cx) - Az\,(Bz\,Cx - Bx\,Cz),$$
$$Az\,(By\,Cz - Bz\,Cy) - Ax\,(Bx\,Cy - By\,Cx),$$
$$Ax\,(Bz\,Cx - Bx\,Cz) - Ay\,(By\,Cz - Bz\,Cy)] \Big]$$

Computational Solution — Sample Problem 2.16

Create vectors **A** and **B** as column matrices with the *z*-component entered as zero.

```
> restart;
> with(linalg):
> A:=[5,3,0]: B:=[3,6,0]:
```

The vectors have been identified and the vector calculations can now be done.

(a)

```
> A+B;
```

$$[8, 9, 0]$$

(b)

```
> dotprod(A,B);
```

$$33$$

```
> magA:=evalf(sqrt(dotprod(A,A)));
```

$$magA := 5.830951895$$

```
> magB:=evalf(sqrt(dotprod(B,B)));
```

$$magB := 6.708203931$$

(c)

```
> theta:=arccos((dotprod(A,B))/(magA*magB))*
  (evalf(180/Pi));
```

$$\theta := 32.47119225$$

(d)

```
> crossprod(A,B);
```

$$\begin{bmatrix} 0 & 0 & 21 \end{bmatrix}$$

(e)

```
> crossprod(B,A);
```

$$\begin{bmatrix} 0 & 0 & -21 \end{bmatrix}$$

Computational Solution — Sample Problem 2.17

Maple can be used to solve this problem, minimizing the need for "by-hand" calculations. First, specify vectors from the origin to the points A, B, and C.

```
> restart;
> with(linalg):
> rA:=[4,0,2]: rB:=[0,2,4]: rC:=[1,3,0]:
```

Scalar or Dot Product

Now we can identify vectors from B to A and from B to C.

```
> rAB:=rA-rB;
```
$$rAB := [4, -2, -2]$$

```
> rCB:=rC-rB;
```
$$rCB := [1, 1, -4]$$

A unit vector normal to the plane is:

```
> nn:=crossprod(rAB,rCB):
> n:=evalm(nn/evalf(sqrt(dotprod(nn,nn))));
```
$$n := \begin{bmatrix} 0.5488212999 & 0.7683498199 & 0.3292927799 \end{bmatrix}$$

The weight is directed downward and in vector notation is:

```
> W:=-50*9.81*[0,0,1]:
```

The normal and tangential components of this vector are:

```
> Wn:=evalm(dotprod(W,n)*n);
```
$$Wn := \begin{bmatrix} -88.64457826 & -124.1024096 & -53.18674695 \end{bmatrix}$$

```
> Wt:=evalm(W-Wn);
```
$$Wt := \begin{bmatrix} 88.64457826 & 124.1024096 & -437.3132530 \end{bmatrix}$$

Computational Solution — Sample Problem 2.18

Maple offers an easy solution to this problem. First, define the three unit vectors and the vector **D** (*d* is used because symbol *D* is protected by *Maple*).

```
> restart;
> with(linalg):
> a:=[-0.231,-0.308,0.923]: b:=[-0.231,0.308,0.923]:
> c:=[0.385,0,0.923]: d:=[0,0,100]:
```

Form the three biorthogonal vectors:

```
> rr:=crossprod(b,c):
> magrr:=sqrt(dotprod(rr,rr));
> r:=evalm(rr/magrr);
```
$$magrr := 0.6466437812$$

$$r := \begin{bmatrix} 0.4396299914 & 0.8792599828 & -.1833776237 \end{bmatrix}$$

```
> ss:=crossprod(c,a):
> magss:=sqrt(dotprod(ss,ss));
> s:=evalm(ss/magss);
```
$$magss := 0.6466437812$$

$$s := \begin{bmatrix} 0.4396299914 & -.8792599828 & -.1833776237 \end{bmatrix}$$

```
> tt:=crossprod(a,b):
> magtt:=sqrt(dotprod(tt,tt));
> t:=evalm(tt/magtt);
```
$$magtt := 0.5861038493$$

$$t := \begin{bmatrix} -.9700806449 & 0. & -.2427829133 \end{bmatrix}$$

```
> A:=dotprod(r,d)/dotprod(r,a);
```
$$A := 33.85698808$$

```
> B:=dotprod(s,d)/dotprod(s,b);
```
$$B := 33.85698808$$

```
> C:=dotprod(t,d)/dotprod(t,c);
```
$$C := 40.62838570$$

3 Particle Equilibirum

Parametric Solutions

Although specific dimensions are given in most undergraduate mechanics problems, greater insight into the problems can often be obtained by seeking a general or parametric solution. An example of this is shown in the solution of Sample Problem 3.4.

Computational Solution — Sample Problem 3.4

This problem will be solved for a range of angles from 0 to 90 degrees. All vectors will be entered as column matrices. Note that although the problem is two dimensional all three components of each vector are designated.

```
> restart;
> with(linalg):
> W:=[0,-50,0]:
> F(theta):=[50*cos(theta),-50*sin(theta),0]:
> T(theta):=-W-F(theta):
> M(theta):=(sqrt(dotprod(T(theta),T(theta)))):
```

We can now plot the magnitude of **T** as the angle changes in radians.

```
> plot(M(theta),theta=0..100*(evalf(Pi/180)),y=60..120,
  labels=[`theta (radians)`,`M(theta)`],color=black);
```

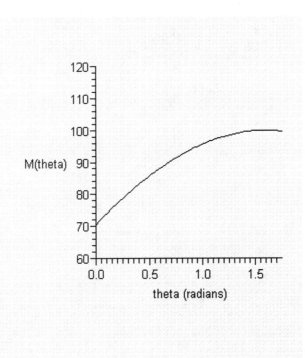

The values of the tensions for any angle may be obtained.

```
> evalf(subs(theta=0,M(theta)));
                    70.71067812

> evalf(subs(theta=90*(evalf(Pi)/180),M(theta)));
                    100.00
```

The angle the cable to the pulley makes with the vertical in radians is:

```
> alpha(theta):=arctan((-50*cos(theta))/(50*
  sin(theta)+50)):
> plot(alpha(theta),theta=0..100*(evalf(Pi/180)),
  y=-60*(evalf(Pi/180))..0,labels=[`theta (radians)`,
  `alpha(theta)`],color=black);
```

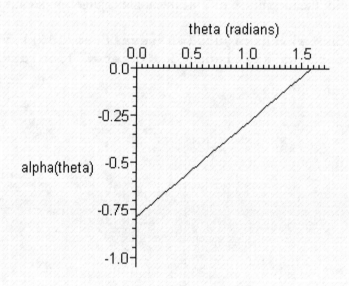

Parametric Solutions

Computational Solution — Sample Problem 3.5

The full vector capabilities of *Maple* are used to solve Sample Problem 3.5. The three simultaneous equations are defined using matrix notation. The solution of the vector equilibrium equation $\mathbf{T_a} + \mathbf{T_b} + \mathbf{T_c} + \mathbf{W} = 0$ is determined, where the magnitudes of the three tensions are unknown.

The unit vectors along the cables are computed by:

```
> restart;
> with(linalg):
> a:=[-4,4,12]: b:=[-4,-6,12]: c:=[5,0,12]:
> au:=evalm(a/sqrt(dotprod(a,a))):
> bu:=evalm(b/sqrt(dotprod(b,b))):
> cu:=evalm(c/sqrt(dotprod(c,c))):
```

The coefficient matrix is:

```
> c:=matrix(3,3,[au[1],bu[1],cu[1],au[2],bu[2],cu[2],au[3],
  bu[3],cu[3]]):
```

The weight can be written as a vector:

```
> w:=matrix(3,1,[0,0,-200*9.81]):
```

The system of equations is $[c][t] = -[w]$. Tensions are in newtons.

```
> matrix(3,1,[ta,tb,tc])=evalf(multiply(inverse(c),-w));
```

$$\begin{bmatrix} ta \\ tb \\ tc \end{bmatrix} = \begin{bmatrix} 723.0242042 \\ 508.6666667 \\ 944.6666667 \end{bmatrix}$$

Computational Solution — Sample Problem 3.6

This problem can be solved using a direct vector solution even though it is in two dimensions by introducing a unit vector **k** in the *z*-direction.

Treat the weight as a variable and then the solution for any weight may be obtained.

```
> restart;
> with(linalg):
> assume(W,real):
> a:=[-sin(25*Pi/180),cos(25*Pi/180),0]:
> b:=[-sin(60*Pi/180),cos(60*Pi/180),0]:
```

```
> t:=[sin(12*Pi/180),cos(12*Pi/180),0]:
> w:=W*[0,1,0]:
> k:=[0,0,1]:
> r:=crossprod(b,k): s:=crossprod(a,k): tt:=crossprod(k,t):
> A:=evalf(dotprod(tt,w)/dotprod(tt,a));
```
$$A := 0.3454744113\ W$$

```
> Ta:=evalf(dotprod(s,w)/dotprod(s,t));
```
$$Ta := 0.7022394680\ W$$

```
> B:=evalf(dotprod(tt,w)/dotprod(tt,b));
```
$$B := 0.2186112889\ W$$

```
> Tb:=evalf(dotprod(r,w)/dotprod(r,t));
```
$$Tb := 0.9105929975\ W$$

Computational Solution — Sample Problem 3.8

```
> restart;
> with(linalg):
> W:=[0,-10*9.81,0]:
> A:=[0.8,0.6,0]: B:=[0,0.8,1.0]:
> t:=(B-A)/sqrt(dotprod(B-A,B-A));
> j:=[0,1,0]:
```
$$t := [-.6172133997, 0.1543033499, 0.7715167496]$$

```
> T:=-dotprod(W,j)/dotprod(t,j);
```
$$T := 635.7606628$$

```
> NN:=-(T*t+W);
```
$$NN := [392.4000001, 0., -490.5000001]$$

Solution of Nonlinear Algebraic Equations

Maple uses the *fsolve* function to numerically solve systems of nonlinear algebraic equations. The *fsolve* function uses an iterative technique. This means the function starts with initial guesses for the unknowns and then iterates until convergence is achieved. The procedure for solving such systems can be outlined as follows.

1. Define all constants.

2. Assign variable names to all equations (this step is optional but clarifies the procedure significantly).

3. Type *fsolve({equation1,equation2, ...},{unknown1,unknown2,...},{unknown1=range1, unknown2=range2,...})*.

Solution of Nonlinear Algebraic Equations

The third set of arguments used with *fsolve* is a set of ranges, one assigned to each unknown, where the solution is to be sought. It should be noted that the algorithm used by *Maple* to solve systems of nonlinear algebraic equations is sensitive to these ranges. The specification of the ranges is optional but is essential when the equations are numerous and/or difficult to solve. For some sets of ranges, the algorithm may not find a solution. For other sets of ranges, the algorithm may find a solution that is not the desired one (the system may have multiple solutions). When using *Maple* to solve these systems, a reasonable estimate of the ranges is necessary to produce the correct solution. Any solution should be compared to expectations before being accepted as correct. Often, there are ways to obtain reasonable guesses for the unknowns.

Computational Solution — Sample Problem 3.9

The weights of the two collars are known.

```
> restart;
> with(linalg):
> WA:=8*9.81: WB:=4*9.81:
```

We can solve the four equilibrium equations using the *fsolve* function.

```
> eq1:=T*cos(alpha)-NA*cos(30*Pi/180)=0:
> eq2:=T*sin(alpha)+NA*sin(30*Pi/180)=WA:
> eq3:=-T*cos(alpha)+NB*sin(45*Pi/180)=0:
> eq4:=-T*sin(alpha)+NB*cos(45*Pi/180)=WB:
```

Initial guesses for the tension, normal forces and the angle are included in the *fsolve* function statement.

```
> fsolve({eq1,eq2,eq3,eq4},{T=0..100,alpha=0..1,NA=0..200,
  NB=0..200});
```

$$\{NB = 105.5448646, NA = 86.17702105, T = 82.59792218, \alpha = 0.4428089395\}$$

```
> evalf(0.442809*180/Pi);
```

$$25.37108682$$

Computational Solution — Sample Problem 3.10

This problem involves three nonlinear equations in terms of the tension T and the two distances a and b.

These equations can be solved using the *fsolve* function.

```
> restart;
> eq1:=T*(0.707*0.5*(0.929*b-0.707*a)+0.707*0.5*(5-
  0.707*a))-10*9.81*0.707=0:
> eq2:=T*(0.371*0.5*(2-0.371*b)-0.929*0.5*(0.929*b-
  0.707*a))-15*9.81*0.929=0:
> eq3:=(2-0.371*b)^2+(0.929*b-0.707*a)^2+(5-0.707*a)^2=4:
> fsolve({eq1,eq2,eq3},{T=0..500,a=0..10,b=0..10});
         {T = 251.3612894, a = 4.847419878, b = 2.836169527}
```

Computational Solution — Sample Problem 3.13

We can obtain guesses for this particular problem by solving the equilibrium equations for the undeformed configuration. This yields:

T_1 = 518 lb, T_2 = 732 lb   (tensions)
δ_1 = 0.259 in., δ_2 = 0.732 in.   (spring deflections)

These deflections are small, which indicates that the geometry of the spring system does not change greatly and the tensions originally obtained are good approximations. This is true of many mechanical systems where the deformation of the members is ignored when considering static equilibrium. The process in small deformation analysis in mechanics of materials courses is to calculate the forces using undeformed geometry and then calculate the stresses and deformations of the part. We will solve Sample Problem 3.13 with stiff springs and show that the answer ignoring deformation is within 5% of the answer when deformation is considered. We will solve the same problem with springs with lower spring constants to show the effect of deformation. The new solution is easily obtained using the same group of cells by changing the values of the spring constants and re-evaluating all the input cells.

```
> restart;
```

Define the known constants.

```
> d:=15.6: L1:=14: L2:=11.4: F:=1000: k1:=2000: k2:=1000:
```

Define the equations.

```
> eq1:=(cos(alpha)-(d^2+(L1+delta1)^2-
```

Solution of Nonlinear Algebraic Equations

```
       (L2+delta2)^2)/(2*d*(L1+delta1)))=0:
>  eq2:=(cos(beta)-(d^2+(L2+delta2)^2-
       (L1+delta1)^2)/(2*d*(L2+delta2)))=0:
>  eq3:=T1-k1*delta1=0:
>  eq4:=T2-k2*delta2=0:
>  eq5:=T1*sin(alpha)+T2*sin(beta)-F=0:
>  eq6:=T1*cos(alpha)-T2*cos(beta)=0:
```

Solve the system using the solution ranges for the unknowns (note that ranges and solutions for α and β are in radians).

```
>  fsolve({eq1,eq2,eq3,eq4,eq5,eq6},{alpha,beta,delta1,
   delta2,T1,T2},{alpha=0..2,beta=0..2,delta1=0..10,
   delta2=0..10,T1=0..1000,T2=0..1000});
```

$$\{T1 = 519.4111966,\ T2 = 709.0933598,\ \alpha = 0.8305635438,$$

$$\beta = 1.054063591,\ \delta 1 = 0.2597055983,\ \delta 2 = 0.7090933598\}$$

These solutions are in agreement with the estimates made for the various unknowns. It can therefore be concluded that this is the correct solution. The small variation between the estimates of T_1, T_2, α, and β and the solutions indicate that the system is very stiff and the deformed geometry is approximately the same as the undeformed geometry.

Computational Solution — Sample Problem 3.13 — Soft Springs

The spring constants used in Sample Problem 3.13 were very stiff, and the solution closely resembled the solution of the springs treated as if they were rigid. It is useful to see the deformation and the tensions if the stiffness of the system is reduced by a factor of 10. This solution is obtained by changing the values for the constants k_1 and k_2 to 200 lb/ft and 100 lb/ft, respectively.

```
>  restart;
```

Define the known constants.

```
>  d:=15.6: L1:=14: L2:=11.4: F:=1000: k1:=200: k2:=100:
```

Define the equations.

```
>  eq1:=(cos(alpha)-(d^2+(L1+delta1)^2-
       (L2+delta2)^2)/(2*d*(L1+delta1)))=0:
>  eq2:=(cos(beta)-(d^2+(L2+delta2)^2-
       (L1+delta1)^2)/(2*d*(L2+delta2)))=0:
>  eq3:=T1-k1*delta1=0:
>  eq4:=T2-k2*delta2=0:
>  eq5:=T1*sin(alpha)+T2*sin(beta)-F=0:
>  eq6:=T1*cos(alpha)-T2*cos(beta)=0:
```

Solve the system using the solution ranges for the unknowns (note that ranges and solutions for α and β are in radians).

```
> fsolve({eq1,eq2,eq3,eq4,eq5,eq6},{alpha,beta,delta1,
    delta2,T1,T2},{alpha=0..2,beta=0..2,delta1=0..10,
    delta2=0..10,T1=0..1000,T2=0..1000});
```

$$\{T2 = 556.5548313,\ T1 = 570.4959470,\ \alpha = 1.097786611,$$
$$\beta = 1.084924926,\ \delta 1 = 2.852479735,\ \delta 2 = 5.565548313\}$$

The system in this case is soft, the tensions are almost equal, and the deformed geometry is approximately an equilateral triangle.

Computational Solution — Sample Problem 3.14

Sample Problem 3.14 is a three-dimensional configuration of deformable cables which can also be solved using *fsolve*. There are nine nonlinear equations in this case.

```
> restart;
```

The coordinates of the attachment point of each cable are:

```
> xA:=-4: yA:=4: zA:=12:
> xB:=-4: yB:=-6: zB:=12:
> xC:=5: yC:=0: zC:=12:
> EAA:=1.2*10^5: EAB:=1.2*10^5: EAC:=1.2*10^5:
```

The unstretched lengths of the cables are:

```
> LA0:=sqrt(xA^2+yA^2+zA^2):
> LB0:=sqrt(xB^2+yB^2+zB^2):
> LC0:=sqrt(xC^2+yC^2+zC^2):
```

The spring constants of the cables are:

```
> kA:=EAA/LA0:
> kB:=EAB/LB0:
> kC:=EAC/LC0:
```

Define the equations.

```
> eq1:=(xA-x0)^2+(yA-y0)^2+(zA-z0)^2=LA^2:
> eq2:=(xB-x0)^2+(yB-y0)^2+(zB-z0)^2=LB^2:
> eq3:=(xC-x0)^2+(yC-y0)^2+(zC-z0)^2=LC^2:
> eq4:=TA=kA*(LA-LA0):
```

Solution of Nonlinear Algebraic Equations

```
> eq5:=TB=kB*(LB-LB0):
> eq6:=TC=kC*(LC-LC0):
> eq7:=((xA-x0)/LA)*TA+((xB-x0)/LB)*TB+((xC-x0)/LC)*TC=0:
> eq8:=((yA-y0)/LA)*TA+((yB-y0)/LB)*TB+((yC-y0)/LC)*TC=0:
> eq9:=((zA-z0)/LA)*TA+((zB-z0)/LB)*TB+((zC-z0)/LC)*TC=1962:
> fsolve({eq1,eq2,eq3,eq4,eq5,eq6,eq7,eq8,eq9},{TA,TB,TC,
  x0,y0,z0,LA,LB,LC},{TA=0..1000,TB=0..1000,TC=0..1000,
  x0=-1/10..1/10,y0=-1/10..1/10,z0=-1/10...1/10,
  LA = 12..16,LB=12..16,LC=12..16});
```

$$\{LA = 13.34638290, LB = 14.06017602, LC = 13.10135064,$$

$$TA = 722.5756001, TB = 515.7944230, TC = 935.5443385,$$

$$x0 = -0.03857814942, y0 = -0.02186934880,$$

$$z0 = -0.09370251247\}$$

A design variation is to choose the spring constants in cables B and C such that the scoreboard would move only downward when it deforms, that is, $x0$ and $y0$ would be zero.

Computational Solution — Sample Problem 3.14 — Vertical Deformation Only

```
> restart;
```

The coordinates of the attachment point of each cable are:

```
> xA:=-4: yA:=4: zA:=12:
> xB:=-4: yB:=-6: zB:=12:
> xC:=5: yC:=0: zC:=12:
> EAA:=1.2*10^5: EAB:=1.2*10^5: EAC:=1.2*10^5:
```

The unstretched lengths of the cables are:

```
> LA0:=sqrt(xA^2+yA^2+zA^2):
> LB0:=sqrt(xB^2+yB^2+zB^2):
> LC0:=sqrt(xC^2+yC^2+zC^2):
```

The spring constant of cable A is:

```
> kA:=EAA/LA0:
```

To restrict deformation to downward motion:

```
> x0:=0:  y0:=0:
```

Define the equations and solve.

```
> eq1:=(xA-x0)^2+(yA-y0)^2+(zA-z0)^2=LA^2:
> eq2:=(xB-x0)^2+(yB-y0)^2+(zB-z0)^2=LB^2:
> eq3:=(xC-x0)^2+(yC-y0)^2+(zC-z0)^2=LC^2:
> eq4:=TA=kA*(LA-LA0):
> eq5:=TB=kB*(LB-LB0):
> eq6:=TC=kC*(LC-LC0):
> eq7:=((xA-x0)/LA)*TA+((xB-x0)/LB)*TB+((xC-x0)/LC)*TC=0:
> eq8:=((yA-y0)/LA)*TA+((yB-y0)/LB)*TB+((yC-y0)/LC)*TC=0:
> eq9:=((zA-z0)/LA)*TA+((zB-z0)/LB)*TB+((zC-z0)/LC)*TC=
  1962:
> fsolve({eq1,eq2,eq3,eq4,eq5,eq6,eq7,eq8,eq9},{TA,TB,TC,
  kB,kC,z0,LA,LB,LC},{TA=0..1000,TB=0..1000,TC=0..1000,
  kB=0..20000,kC=0..20000,z0=-1/10...1/10,
  LA = 12..16,LB=12..16,LC=12..16});
```

$$\{LA = 13.34632678, LB = 14.07566832, LC = 13.08145399,$$
$$TA = 722.0679602, TB = 507.6847124, TC = 943.6502848,$$
$$kB = 6709.343298, kC = 11585.07099, z0 = -0.08819418097\}$$

The required *EA* values for each cable are the following.

```
> kB:=6709.343298: kC:=11585.07099:
> kALA0:=kA*LA0; kBLB0:=kB*LB0; kCLC0:=kC*LC0;
```

$$kALA0 := 1.200000000 \; 10^5$$

$$kBLB0 := 93930.80617$$

$$kCLC0 := 1.506059229 \; 10^5$$

Notice that to meet this condition it was necessary to decrease the stiffness of cable *B* and increase the stiffness of cable *C*. This could be obtained either by changing the cross-sectional area of the cables or by changing the material used in the cables.

Computational Solution — Sample Problem 3.15

The statically indeterminate Sample Problem 3.15 which involves 11 equations in 11 unknowns is also solved using the *fsolve* function.

```
> restart;
```

Solution of Nonlinear Algebraic Equations

The coordinates of the attachment point of each cable are:

```
> xA:=-4:  yA:=4:   zA:=12:  EAA:=1.2*10^5:
> xB:=-4:  yB:=-6:  zB:=12:  EAB:=1.2*10^5:
> xC:=5:   yC:=0:   zC:=12:  EAC:=1.2*10^5:
> xD:=0:   yD:=0:   zD:=12:  EAD:=12*10^5:
```

The unstretched lengths of the cables are:

```
> LA0:=sqrt(xA^2+yA^2+zA^2):
> LB0:=sqrt(xB^2+yB^2+zB^2):
> LC0:=sqrt(xC^2+yC^2+zC^2):
> LD0:=sqrt(xD^2+yD^2+zD^2):
```

The spring constants of the cables are:

```
> kA:=EAA/LA0:
> kB:=EAB/LB0:
> kC:=EAC/LC0:
> kD:=EAD/LD0:
```

Define the equations and solve.

```
> eq1:=(xA-x0)^2+(yA-y0)^2+(zA-z0)^2=LA^2:
> eq2:=(xB-x0)^2+(yB-y0)^2+(zB-z0)^2=LB^2:
> eq3:=(xC-x0)^2+(yC-y0)^2+(zC-z0)^2=LC^2:
> eq4:=(xD-x0)^2+(yD-y0)^2+(zD-z0)^2=LD^2:
> eq5:=TA=kA*(LA-LA0):
> eq6:=TB=kB*(LB-LB0):
> eq7:=TC=kC*(LC-LC0):
> eq8:=TD=kD*(LD-LD0):
> eq9:=((xA-x0)/LA)*TA+((xB-x0)/LB)*TB+((xC-
  x0)/LC)*TC+((xD-x0)/LD)*TD=0:
> eq10:=((yA-y0)/LA)*TA+((yB-y0)/LB)*TB+((yC-
  y0)/LC)*TC+((yD-y0)/LD)*TD=0:
> eq11:=((zA-z0)/LA)*TA+((zB-z0)/LB)*TB+((zC-
  z0)/LC)*TC+((zD-z0)/LD)*TD=1962:
> fsolve({eq1,eq2,eq3,eq4,eq5,eq6,eq7,eq8,eq9,eq10,eq11},
  {TA,TB,TC,TD,x0,y0,z0,LA,LB,LC,LD},{TA=0..5000,
  TB=0..5000,TC=0..5000,TD=0..5000,x0=-1/10..1/10,
  y0=-1/10..1/10,z0=-1/10..1/10,LA=12..16,LB=12..16,
  LC=12..16,LD=12..16});
```

$\{LA = 13.28032330,\ LB = 14.01042696,\ LC = 13.01751581,$
$LD = 12.01622958,\ TA = 125.0440451,\ TB = 89.37396927,$
$TC = 161.6843960,\ TD = 1622.958050,$
$x0 = -0.006594451270,\ y0 = -0.003746167801,$
$z0 = -0.01622718705\}$

Notice that the addition of the fourth cable in the position chosen causes that cable to carry almost all the weight of the scoreboard and the other cables only stabilize the system. A more reasonable design is obtained by attaching the fourth cable at $xD = 4$ and $yD = -4$.

Computational Solution — Sample Problem 3.15 — Design Variation

```
> restart;
```

The coordinates of the attachment point of each cable are:

```
> xA:=-4: yA:=4: zA:=12: EAA:=1.2*10^5:
> xB:=-4: yB:=-6: zB:=12: EAB:=1.2*10^5:
> xC:=5: yC:=0: zC:=12: EAC:=1.2*10^5:
> xD:=4: yD:=-4: zD:=12: EAD:=12*10^5:
```

The unstretched lengths of the cables are:

```
> LA0:=sqrt(xA^2+yA^2+zA^2):
> LB0:=sqrt(xB^2+yB^2+zB^2):
> LC0:=sqrt(xC^2+yC^2+zC^2):
> LD0:=sqrt(xD^2+yD^2+zD^2):
```

The spring constants of the cables are:

```
> kA:=EAA/LA0:
> kB:=EAB/LB0:
> kC:=EAC/LC0:
> kD:=EAD/LD0:
```

Define the equations and solve.

```
> eq1:=(xA-x0)^2+(yA-y0)^2+(zA-z0)^2=LA^2:
> eq2:=(xB-x0)^2+(yB-y0)^2+(zB-z0)^2=LB^2:
> eq3:=(xC-x0)^2+(yC-y0)^2+(zC-z0)^2=LC^2:
> eq4:=(xD-x0)^2+(yD-y0)^2+(zD-z0)^2=LD^2:
> eq5:=TA=kA*(LA-LA0):
```

Solution of Nonlinear Algebraic Equations

```
> eq6:=TB=kB*(LB-LB0):
> eq7:=TC=kC*(LC-LC0):
> eq8:=TD=kD*(LD-LD0):
> eq9:=((xA-x0)/LA)*TA+((xB-x0)/LB)*TB+((xC-
  x0)/LC)*TC+((xD-x0)/LD)*TD=0:
> eq10:=((yA-y0)/LA)*TA+((yB-y0)/LB)*TB+((yC-
  y0)/LC)*TC+((yD-y0)/LD)*TD=0:
> eq11:=((zA-z0)/LA)*TA+((zB-z0)/LB)*TB+((zC-
  z0)/LC)*TC+((zD-z0)/LD)*TD=1962:
> fsolve({eq1,eq2,eq3,eq4,eq5,eq6,eq7,eq8,eq9,eq10,eq11},
  {TA,TB,TC,TD,x0,y0,z0,LA,LB,LC,LD},{TA=0..2000,
  TB=0..2000,TC=0..2000,TD=0..2000,x0=-1/5..1/5,
  y0=-1/5..1/5,z0=-1/5..1/5,LA=12..16,LB=12..16,LC=12..16,
  LD=12..16});
```

$\{LA = 13.36956812, LB = 14.02250147, LC = 13.03641933,$

$LD = 13.27433600, TA = 932.2937864, TB = 192.8697369,$

$TC = 336.1784261, TD = 708.8683804, x0 = 0.05284450610,$

$y0 = -.1057402133, z0 = -0.06079186672\}$

When this position of attachment is chosen for the fourth cable, all the tensions are less than 1000 N.

4 Rigid Bodies: Equivalent Force Systems

Chapter 4 considers rigid bodies where the point of application of the force on the body is important to the equilibrium of the rigid body. The turning effect of a force is defined as the moment of the force about a certain point. The moment is defined as the cross product (vector product) of the position vector from the point to a point on the line of action of the force with the force vector: **M** = **r** × **F**. In this chapter of the supplement, we will show how *Maple* can be used to numerically and symbolically solve problems using cross products. Those sample problems where *Maple* is useful will be solved in detail.

Computational Solution — Sample Problem 4.4

```
> restart;
> with(linalg):
> P:=[500,0,0]: rBA:=[0,0,-4]: rBC:=[0,2,-4]:
> MA:=crossprod(rBA,P); MC:=crossprod(rBC,P);
```

$$MA := \begin{bmatrix} 0 & -2000 & 0 \end{bmatrix}$$

$$MC := \begin{bmatrix} 0 & -2000 & -1000 \end{bmatrix}$$

```
> magMA:=evalf(sqrt(dotprod(MA,MA)));
> magMC:=evalf(sqrt(dotprod(MC,MC)));
```

$$magMA := 2000.$$

$$magMC := 2236.067977$$

Computational Solution — Sample Problem 4.5

The direct vector solution is easily obtained using *Maple* as a vector calculator.

```
> restart;
> with(linalg):
```

Define the moment $\mathbf{M}_0$ and the position vector $\mathbf{r}_{A/0}$.

```
> M0:=[3000,1000,-3000]: rA0:=[10,-6,8]:
> Fp:=evalm(-(crossprod(rA0,M0))/(dotprod(rA0,rA0)));
```

$$Fp := \begin{bmatrix} -50 & -270 & -140 \end{bmatrix}$$

Computational Solution — Sample Problem 4.6

Although Varignon's theorem can be used to find the moment in many cases, especially when the dimensions are given in such a manner that perpendicular distances are obvious, vector algebra is usually the best way to write the moment equations:

```
> restart;
> with(linalg):
```

The moment in ft lb is:

```
> M0:=crossprod([4,2,0],[70.7,70.7,0]);
```
$$M0 := \begin{bmatrix} 0. & 0. & 141.4 \end{bmatrix}$$

Computational Solution — Sample Problem 4.7

This problem can be solved by hand or using a calculator but the numerical work is easily done using *Maple*.

```
> restart;
> with(linalg):
> rba:=[3,4,0]: rca:=[0,2,4]: rcb:=[-3,-2,4]:
> t:=evalf(rcb/sqrt(dotprod(rcb,rcb)));
```
$$t := [-.5570860145, -.3713906762, 0.7427813527]$$

```
> T:=300*t;
```
$$T := [-167.1258044, -111.4172029, 222.8344058]$$

```
> Ma:=crossprod(rba,T);
```
$$Ma := \begin{bmatrix} 891.3376232 & -668.5032174 & 334.2516089 \end{bmatrix}$$

```
> Mb:=crossprod(rca,T);
```
$$Mb := \begin{bmatrix} 891.3376232 & -668.5032176 & 334.2516088 \end{bmatrix}$$

```
> p=evalm(crossprod(T,Ma)/dotprod(T,T));
```
$$p = \begin{bmatrix} 1.241379310 & 2.827586206 & 2.344827586 \end{bmatrix}$$

Computational Solution — Sample Problem 4.8

This problem does not need to be solved using computational software, but it may be used to solve a system of two equations in two unknowns.

```
> restart;
> with(linalg):
```

```
> c:=[[-8.018,16.04],[16.64,0]];
```
$$c := [[-8.018, 16.04], [16.64, 0]]$$

```
> Mab:=[0,-33.42]:
> multiply(-inverse(c),Mab);
```
$$\begin{bmatrix} 2.008413462 & 1.003956305 \end{bmatrix}$$

Computational Solution — Sample Problem 4.9

This problem has been reduced to the solution of two equations (one nonlinear). We will solve these using the *fsolve* function.

```
> restart;
> eq1:=0.686*Tb+0.766*Tc=3.6:
> eq2:=5.89*Tb^2-1.77*Tb*Tc+13.397*Tc^2=100:
> fsolve({eq1,eq2});
```
$$\{Tc = 1.241129241, Tb = 3.861946066\}$$

Computational Solution — Sample Problem 4.10

The component of a vector along an axis in space is an example of the use of the scalar triple product.

The moment in ft lb is computed as follows. Note that it is sometimes necessary to use the operator *evalm* to get expressions involving vectors and matrices to simplify fully.

```
> restart;
> with(linalg):
> i:=[1,0,0]:
> rBA:=[sin(30*evalf(Pi/180)),cos(30*evalf(Pi/180)),0]*
  (15/12):
> F:=[0,0,50]:
> Mx:=(dotprod(i,crossprod(rBA,F)))*i;
```
$$Mx := [54.12658775, 0, 0]$$

Computational Solution — Sample Problem 4.11

```
> restart;
> with(linalg):
> rrA:=[-.9*sin(20*evalf(Pi/180)),.9*
  cos(20*evalf(Pi/180)),.4]:
> Fr:=[-80,-80,0]:
> rlA:=[-.9*sin(20*evalf(Pi/180)),.9*
  cos(20*evalf(Pi/180)),-.4]:
> Fl:=[-80,-80,40]:
```

Computational Solution — Sample Problem 4.13

```
> MA:=evalm(crossprod(rrA,Fr)+crossprod(rlA,Fl));
```
$$MA := \begin{bmatrix} 33.82893435 & 12.31272516 & 184.5666380 \end{bmatrix}$$

```
> j:=[0,1,0]:
> My:=dotprod(j,MA);
```
$$My := 12.31272516$$

Sample Problems 4.12 and 4.13 require cross products to determine the applied moments. The vector capabilities of *Maple* reduce the numerical labor in solving these problems.

Computational Solution — Sample Problem 4.12

All moments are in N m.

```
> restart;
> with(linalg):
> rBA:=[-6,10,0]:  rDC:=[-3,-10,2]:
> FB:=[0,0,-8000]:    FD:=[0,-2000,0]:
> CAB:=crossprod(rBA,FB);  CCD:=crossprod(rDC,FD);
```
$$CAB := \begin{bmatrix} -80000 & -48000 & 0 \end{bmatrix}$$

$$CCD := \begin{bmatrix} 4000 & 0 & 6000 \end{bmatrix}$$

```
> M:=evalm(CAB+CCD);
```
$$M := \begin{bmatrix} -76000 & -48000 & 6000 \end{bmatrix}$$

Computational Solution — Sample Problem 4.13

```
> restart;
> with(linalg):
> rAB:=[0,0,2]:
> FA:=[1000,0,0]:
> cB:=crossprod(rAB,FA);
```
$$cB := \begin{bmatrix} 0 & 2000 & 0 \end{bmatrix}$$

```
> rAC:=[-6,0,2]:
> cC:=crossprod(rAC,FA);
```
$$cC := \begin{bmatrix} 0 & 2000 & 0 \end{bmatrix}$$

```
> rAD:=[-6,3,2]:
> cD:=crossprod(rAD,FA);
```
$$cD := \begin{bmatrix} 0 & 2000 & -3000 \end{bmatrix}$$

Maple is very useful for performing the vector calculations necessary to compute equivalent force systems, as shown in the computational solutions of Sample Problems 4.14, 4.16, and 4.17.

Computational Solution — Sample Problem 4.14

Define the force and position vectors of the coplanar force system.

```
> restart;
> with(linalg):
> F1:=[100,100,0]: F2:=[0,50,0]: F3:=[300,-450,0]:
> r1:=[2,3,0]: r2:=[-10,2,0]: r3:=[4,-4,0]:
> R:=F1+F2+F3;
```
$$R := [400, -300, 0]$$

```
> r1:=[2,3,0]: r2:=[-10,2,0]: r3:=[4,-4,0]:
> C:=evalm(crossprod(r1,F1)+crossprod(r2,F2)+
  crossprod(r3,F3));
```
$$C := \begin{bmatrix} 0 & 0 & -1200 \end{bmatrix}$$

```
> r:=evalf(evalm(((crossprod(R,C))/(dotprod(R,R)))));
```
$$r := \begin{bmatrix} 1.440000000 & 1.920000000 & 0. \end{bmatrix}$$

Calculation of the equivalent wrench for a given force system involves many vector operations that are most easily done using *Maple* as shown in Sample Problems 4.16 and 4.17.

Computational Solution — Sample Problem 4.16

```
> restart;
> with(linalg):
```

The force and moment vectors at the instrument center are:

```
> R:=[50,150,800]: M0:=[80,10,10]:
```

The unit vector in the **R**-direction is:

```
> eR:=evalf(normalize(R));
```
$$eR := \begin{bmatrix} 0.06131393394 & 0.1839418019 & 0.9810229432 \end{bmatrix}$$

The torque which is parallel to **R** is:

```
> T:=evalm((dotprod(M0,eR))*eR);
```

Computational Solution — Sample Problem 4.17

$$T := \begin{bmatrix} 1.015037594 & 3.045112784 & 16.24060151 \end{bmatrix}$$

The perpendicular component of the couple is:

> `evalm(M0-T);`

$$\begin{bmatrix} 78.98496241 & 6.954887216 & -6.24060151 \end{bmatrix}$$

The perpendicular vector from the origin to the wrench axis is:

> `p0:=evalm((crossprod(R,(M0-T)))/(dotprod(R,R)));`

$$p0 := \begin{bmatrix} -0.009774436090 & 0.09548872182 & -0.01729323308 \end{bmatrix}$$

Note that this is not the vector to the surface of the plate but a vector perpendicular to the wrench axis.

The intercept with the surface of the force plate **r** can now be found by moving from the perpendicular intercept point along the wrench axis to the surface of the plate a distance d. This is expressed as $\mathbf{r} = \mathbf{p} + d\,\mathbf{ru}$.

> `eq1:=x+0.01-0.061*d:`
> `eq2:=y-0.095-0.184*d:`
> `eq3:=-0.04+0.017-0.981*d:`
> `fsolve({eq1,eq2,eq3});`

$$\{d = -0.02344546382,\ y = 0.09068603466,\ x = -0.01143017329\}$$

Computational Solution — Sample Problem 4.17

Torque is in N m.

> `restart;`
> `with(linalg):`
> `R:=[50,-150,450]: M0:=[13.5,9,-6]: n:=[0,0,1]:`
> `T:=(dotprod(R,M0))/(dotprod(R,n));`

$$T := -7.500000000$$

Position is in meters.

> `p0:=evalm((crossprod(R,(M0-T*n)))/(dotprod(R,R)));`

$$p0 := \begin{bmatrix} -0.01879120879 & 0.02637362637 & 0.01087912088 \end{bmatrix}$$

5 Distributed Forces: Centroids and Center of Gravity

To determine the centroid of a line, area, or volume, integrals must be evaluated. Although this is a standard topic in introductory calculus courses, these integrals can be evaluated either numerically or symbolically using *Maple*. Examples of both symbolic and numerical evaluation are shown in Computational Window 5.1.

Numerical and Symbolic Integration
Integrals may be evaluated either numerically or symbolically using computational software. Dimensions must be specified if the integral is evaluated numerically. Integration is performed using the *int* operator. The first argument of the *int* operator is the integrand and the second argument specifies the limits of integration.

Computational Window 5.1

The centroid of the triangle shown in Figure 5.14 of the *Statics* text determined using single integration is:

```
> restart;
> yc:=(2/h^2)*(int(h*y-y^2,y=0..h));
```

$$yc := \frac{1}{3} h$$

The integral could be evaluated symbolically using double integration. Note that double integration is performed using nested *int* operators with additional arguments specifying variables and limits of integration.

```
> yc:=simplify((2/(b*h))*(int(int(y,x=(a*y/h)..(b-
  ((b-a)/h)*y)),y=0..h)));
```

$$yc := \frac{1}{3} h$$

Note that although computational software replaces tables of integrals, the integrals must be correctly formulated before these tools can be used.

Computational Solution — Sample Problem 5.1

The area and centroid are determined symbolically using *Maple*. The double integrals are set up using multivariable calculus techniques. These integrals are simple enough to be performed by hand.

The centroid of a semicircular area is computed using symbolic manipulations instead of a table of integrals.

Numerical and Symbolic Integration

```
> restart;
> A:=int(int(r,r=0..R),theta=0..Pi);
```
$$A := \frac{1}{2} R^2 \pi$$

```
> y:=(2/(Pi*R^2))*(int((int(r^2,r=0..R))*sin(theta),
  theta=0..Pi));
```
$$y := \frac{4}{3} \frac{R}{\pi}$$

Computational Solution — Sample Problem 5.3

The length and centroid of the curves for $n = 1$ to 5 is obtained by numerical integration. The procedure for numerical integration is very similar to that for symbolic integration. Using the capitalized integration operator *Int* instead of the symbolic integration operator *int* tells *Maple* to bypass the symbolic procedure. This decreases computation time. The operator *evalf* is used to tell *Maple* to return a numerical answer. Iterative calculations can be performed using the *Maple* programming constructs. The following computations are repeated five times each through the use of a *for* loop. Note that the *Maple* term *od* is used to close the loop and must be included.

The length of the curve for $n = 1$ to 5 is:

```
> restart;
> for n from 1 to 5 do
    L(n):=evalf(Int(sqrt(n^2*x^(2*(n-1))+1),x=0..1))
  od;
```

$$L(1) := 1.414213562$$

$$L(2) := 1.478942858$$

$$L(3) := 1.547865655$$

$$L(4) := 1.600229428$$

$$L(5) := 1.640559758$$

The *x*-coordinate of the centroid of the curve for $n = 1$ to 5 is:

```
> for n from 1 to 5 do
    xc(n):=(1/L(n))*evalf(Int((sqrt(n^2*x^(2*
    (n-1))+1))*x,x=0..1))
  od;
```

$$xc(1) := 0.5000000002$$

$xc(2) := 0.5736270693$

$xc(3) := 0.6086488300$

$xc(4) := 0.6306272271$

$xc(5) := 0.6460120515$

The *y*-coordinate of the centroid of the curve for $n = 1$ to 5 is:

```
> for n from 1 to 5 do
  yc(n):=(1/L(n))*evalf(Int((sqrt(n^2*
  x^(2*(n-1))+1))*x^n,x=0..1))
  od;
```

$yc(1) := 0.5000000002$

$yc(2) := 0.4099802173$

$yc(3) := 0.3663680073$

$yc(4) := 0.3417886220$

$yc(5) := 0.3261142115$

For $n = 5$, the curve is:

```
> plot(x^5,x=0..1,y=0..1,labels=[`x`,`x^5`],color=black);
```

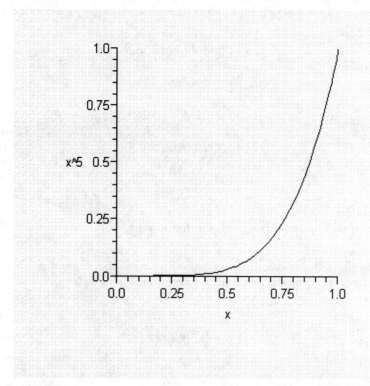

Three-Dimensional Scatter Plots

A valuable graphic tool in *Maple* is the 3D point plotting function *pointplot3d*. The *pointplot3d* function allows you to plot a collection of points in three-dimensional space. The arguments of the *pointplot3d* function are triplets of numbers, one for each point in the plot. The general syntax is of the form *pointplot3d([x1,y1,z1],[x2,y2,z2],...)*. Options can be added as arguments of the *pointplot3D* function to customize the plot. 3D scatter plots are very valuable when looking at centroids of lines in space.

Computational Window 5.2

A helical spiral of radius r and pitch p is plotted using Eq. (5.24) of the *Statics* text. Note that the function *pointplot3d* is located in the *plots* package. It cannot be used until this package is loaded.

```
> restart;
> with(plots):
> R:=2:
> beta:=evalf(Pi/6):
> r:=R*cos(beta):
> for i from 1 to 241 do
    s:=array(1..241);
    x:=array(1..241);
    y:=array(1..241);
    z:=array(1..241);
    xyz:=array(1..241);
  od:
> for i from 1 to 241 do
    s[i]:=(i-1)/10;
    theta[i]:=s[i]/r;
    x[i]:=add(cos(theta[n])*cos(beta)/10,n=1..i);
    y[i]:=-r+add(sin(theta[n])*cos(beta)/10,n=1..i);
    z[i]:=add(sin(beta)/10,n=1..i);
    xyz[i]:=[x[i],y[i],z[i]];
  od:
> pointplot3d(xyz,color=black,axes=box,
  labels=['x','y','z']);
```

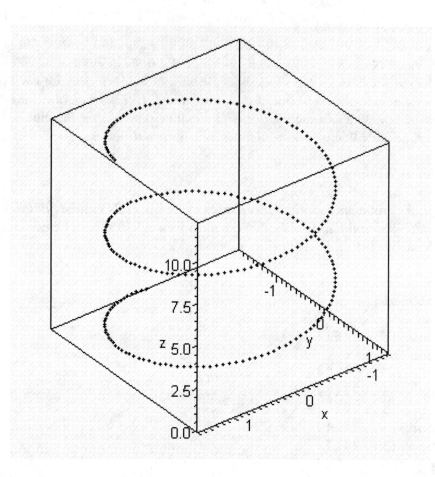

Computational Solution — Sample Problem 5.4

Determine the surface area of the surface of revolution of a half sine curve of amplitude 0.5 m beginning at (0,2) m. An equation for this curve is $y(x) = 2 + 0.5 \sin(\pi x)$ from $x = 0$ to $x = 1$.

Functions are defined in *Maple* using the syntax *f:=x->expression*. This command defines *f* to be the function of *x* given by *expression*. Several additional plotting features are introduced in the solution of this problem. Text can be added to a plot in precise locations using the *textplot* command. This command is located in the *plots* package and must be loaded using the command *with(plots)*. To represent multiple graphics objects on a single set of axes, the graphics objects may be assigned variable names. All graphics objects can then be displayed on the same axes using the *display* command.

```
> restart;
> with(plots):
> y:=x->2+0.5*sin(Pi*x):
> ds:=x->sqrt(1+(0.5*Pi*cos(Pi*x))^2):
> L:=int(ds(x),x=0..1);
```
$$L := 1.463695472$$

```
> yc:=(1/L)*int(y(x)*ds(x),x=0..1);
```
$$yc := 2.287854706$$

```
> plot1:=plot([[s,y(s),s=0..1],[s,yc,s=0..1]],
  x=-0..1,y=2..2.6,labels=[`x`,``],
  labelfont=[TIMES,BOLD,12],color=[black,black],
  linestyle=[SOLID,DOT]):
> textplot1:=textplot([0.5,2.55,"y(x)"],
  labelfont=[TIMES,BOLD,12]):
> textplot2:=textplot([0.5,2.33,"yc"],
  labelfont=[TIMES,BOLD,12]):
> display([plot1,textplot1,textplot2]);
```

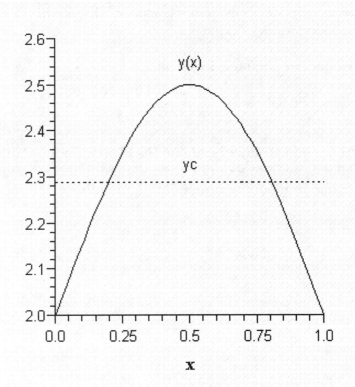

6 Equilibrium of Rigid Bodies

Several sample problems from Chapter 6 are solved using *Maple* as examples of how this software may be used to reduce the computational burden. For example, Sample Problem 6.4 required the solution of a transcendental equation. This problem can be solved using *Maple* by first graphing Eq. (SP6.4.6) and then finding an initial estimate of the root and limiting the domain over which the root is sought using these estimates. The *fsolve* function can then be used to determine the exact root. Although we usually think of roots of quadratic or cubic equations, this operator can be used to solve for the roots of transcendental equations. Since multiple roots may exist, it is necessary to restrict the domain over which the root is sought to produce the root that is the correct solution to the problem. The domain over which the root is sought will be referred to as the search domain from this point forward.

Computational Solution — Sample Problem 6.4

A plot of Eq. (SP6.4.6) in radians is:

```
> restart;
> f:=alpha->(sin(alpha))^2-(cos(alpha))^2+.75*cos(alpha):
> plot(f(alpha),alpha=0..7,y=-2..2,labels=[`alpha
  (radians)`,`f(alpha)`],labelfont=[TIMES,BOLD,12],
  color=black);
```

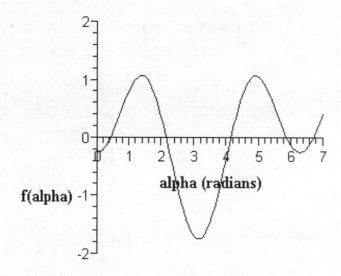

We can see that the smallest positive root of this equation is approximately 0.4 radians or 23 degrees. We will graph the function from 0 to 45 degrees to confirm the estimate of the root so that the search domain can be defined and the *fsolve* function can be used. Note that if we chose the search domain to be between 1 and 3, the *fsolve* function would converge to the second root of the transcendental equation or if we chose the search domain to be between 3 and 5, the *fsolve* function would converge to the third root of the transcendental equation.

```
> plot(f(evalf(alpha*Pi/180)),alpha=0..50,y=-.5..1,
  labels=[`alpha(degrees)`,`f(alpha)`],
  labelfont=[TIMES,BOLD,12],color=black);
```

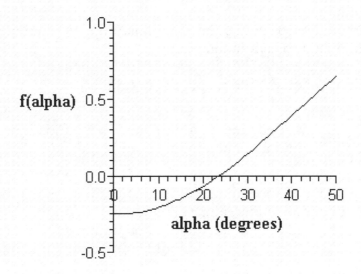

The root of the equation can be estimated from the graph to be approximately 25 degrees. The function *fsolve* can be used to determine the root in degrees more precisely.

```
> fsolve(f(evalf(alpha*Pi/180)),alpha,23..27);
                  23.21331814
```

The system of nonlinear equations in Eqs. (SP6.4.1–3) can be solved directly using *Maple*. Again, as in any iterative technique, an initial estimate of the unknowns must be given.

Note that α is solved for in degrees.

```
> W:=1:
> eq1:=P*cos(evalf(alpha*Pi/180))-
  W*sin(evalf(alpha*Pi/180))=0:
> eq2:=N+P*sin(evalf(alpha*Pi/180))-
  W*cos(evalf(alpha*Pi/180))=0:
> eq3:=P*sin(evalf(alpha*Pi/180))*2*cos(evalf(alpha*
  Pi/180))-W*cos(evalf(alpha*Pi/180))*(2*
  cos(evalf(alpha*Pi/180))-1.5)=0:
> fsolve({eq1,eq2,eq3},{P,N,alpha},{P=0..1,N=0..1,
  alpha=0..25});
```

$$\{\alpha = 23.21331814, N = 0.7500000000, P = 0.4288757202\}$$

Although the angle is independent of the weight of the rod, the forces P and N depend upon the weight. A unit weight was assumed in the solution.

This problem is solved by hand in the text but that does little to increase one's understanding of a rigid body in equilibrium.

Computational Solution — Sample Problem 6.5

```
> restart;
> W:=50*9.81:
> k:=500:
> l:=2:
> T:=k*l*(sqrt(5-4*cos(alpha))-1):
> f:=2*T/sqrt(5-4*cos(alpha))-W/2:
> fsolve(f);
```

$$0.3891423413$$

```
> evalf(0.389142*180/Pi);
```

$$22.29619423$$

```
> plot(f,alpha=0..Pi/5,y=-400..400,labels=[`alpha
  (radians)`,`f(alpha)`],labelfont=[TIMES,BOLD,12],
  color=black);
```

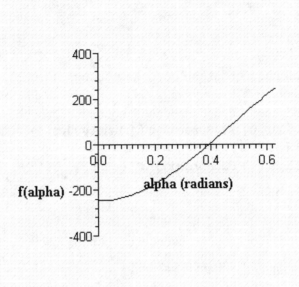

Computational Solution — Sample Problem 6.6

The solution of the system of simultaneous equations in Sample Problem 6.6 is obtained using *Maple*. The unit vectors along each of the cables are obtained using the vector algebra capabilities of *Maple*.

```
> restart;
> with(linalg):
> rBE:=[-300,100,150]: rBF:=[-300,200,0]:
  rCG:=[-300,200,-80]:
> eBE:=evalf(normalize(rBE)); eBF:=evalf(normalize(rBF));
  eCG:=evalf(normalize(rCG));
```

$$eBE := \begin{bmatrix} -.8571428571 & 0.2857142857 & 0.4285714286 \end{bmatrix}$$

$$eBF := \begin{bmatrix} -.8320502943 & 0.5547001960 & 0. \end{bmatrix}$$

$$eCG := \begin{bmatrix} -.8122955414 & 0.5415303611 & -.2166121444 \end{bmatrix}$$

The equilibrium equations may be generated using the *Maple* symbolic processor. The left hand side of the force and moment vector equation is written in symbolic notation and then evaluated symbolically.

Computational Solution — Sample Problem 6.6 — Symbolic Generation of Equilibrium Equations

```
> restart;
> with(linalg):
```

The following symbolic expression for the summation of forces or the resultant of the forces is equal to zero.

```
> vector([0,-1000,0])+vector([Ax,Ay,Az])+TBE*vector(
  [-.857,.286,.429])+TBF*vector(
  [-.832,.555,0])+TCG*vector([-.812,.542,-.217]);
```

$$\begin{bmatrix} 0 & -1000 & 0 \end{bmatrix} + \begin{bmatrix} Ax & Ay & Az \end{bmatrix}$$
$$+ TBE \begin{bmatrix} -.857 & 0.286 & 0.429 \end{bmatrix} + TBF \begin{bmatrix} -.832 & 0.555 & 0 \end{bmatrix}$$
$$+ TCG \begin{bmatrix} -.812 & 0.542 & -.217 \end{bmatrix}$$

The percent sign % can be used to represent the result of the previous *Maple* command. In this case, the function *evalm* is used to combine and simplify the result of the first series of commands.

```
> evalm(%);
```

$$\big[[Ax - 0.857\ TBE - 0.832\ TBF - 0.812\ TCG,$$
$$-1000 + Ay + 0.286\ TBE + 0.555\ TBF + 0.542\ TCG,$$
$$Az + 0.429\ TBE - 0.217\ TCG] \big]$$

The following symbolic expression for the summation of moments about the origin is equal to zero.

```
> (crossprod(vector(3,[.3,0,0]),(TBE*(vector(3,
  [-.857,.286,.429])))+TBF*(vector(3,[-.832,.555,0]))))+
  crossprod(vector(3,[.3,0,-.07]),TCG*(vector(3,
  [-.812,.542,-.217])))+crossprod(vector(3,[.3,0,-.15]),
  vector(3,[0,-1000,0]));
```

$$\begin{bmatrix} 0 & -0.1287\ TBE & 0.0858\ TBE + 0.1665\ TBF \end{bmatrix}$$
$$+ \begin{bmatrix} 0.03794\ TCG & 0.12194\ TCG & 0.1626\ TCG \end{bmatrix}$$
$$+ \begin{bmatrix} -150.00 & -0. & -300.0 \end{bmatrix}$$

> `evalm(%);`

$$[[0.03794\ TCG - 150.00, -0.1287\ TBE + 0.12194\ TCG,$$
$$0.0858\ TBE + 0.1665\ TBF + 0.1626\ TCG - 300.0]]$$

Linear systems of equations can be easily solved using the functions *solve* and *fsolve* for exact and approximate floating point results, respectively. However, linear systems may also be solved manually using the techniques of linear algebra. The coefficient matrix for the system of six linear equations can be formed and the equations solved.

Computational Solution — Sample Problem 6.6 — Solution of Six Equations of Equilibrium

The six scalar equations are written in matrix notation as follows. Matrices can be defined in *Maple* using the function *matrix(arg1,arg2,arg3)* where argument one is the number of rows, argument two is the number of columns, and argument three is a list of the elements proceeding along the rows from the upper left to the lower right.

> `restart;`
> `with(linalg):`
> `C:=matrix(6,6,[1,0,0,-.857,-.832,-.812,0,1,0,.286,`
  `.555,.542,0,0,1,.429,0,-.217,0,0,0,0,0,37.907,0,0,0,`
  `-128.571,0,121.844,0,0,0,85.714,166.41,162.429]);`

$$C := \begin{bmatrix} 1 & 0 & 0 & -.857 & -.832 & -.812 \\ 0 & 1 & 0 & 0.286 & 0.555 & 0.542 \\ 0 & 0 & 1 & 0.429 & 0 & -.217 \\ 0 & 0 & 0 & 0 & 0 & 37.907 \\ 0 & 0 & 0 & -128.571 & 0 & 121.844 \\ 0 & 0 & 0 & 85.714 & 166.41 & 162.429 \end{bmatrix}$$

> `P:=1000*matrix(6,1,[0,1,0,150,0,300]);`

$$P := 1000 \begin{bmatrix} 0 \\ 1 \\ 0 \\ 150 \\ 0 \\ 300 \end{bmatrix}$$

> `[Ax,Ay,Az,TBC,TBF,TCG]=transpose(multiply(inverse(C),P));`

$$([Ax, Ay, Az, TBC, TBF, TCG]) = [3106.244582, -2.1331765,$$
$$-750.0758508, 3750.014698, -3991.159708, 3957.052786]$$

The reactions are (notice that the rod *BF* is in compression.):

A = 3106**i** − 2.133**j** − 750.1**k**

TBE = 3750(−.857**i** + .286**j** + .429**k**)

TBF = −3991(−.832**i** + .555**j**)

TCG = 3957(−.812**i** + .542**j** + −.217**k**)

Computational Solution — Sample Problem 6.7 — Symbolic Generation and Solution of Equilibrium Equations

Equilibrium equations can be generated using the symbolic processor. Symbolic solutions to systems of nonlinear equations can be obtained using the *solve* function.

The resultant of the forces is zero.

```
> restart;
> with(plots):
> with(linalg):
> P*vector([0,cos(theta),-sin(theta)])+vector([Ax,Ay,Az])+
  vector([0,-100,0])+vector([0,By,Bz]);
```

$$P\,[\,0\;\;\cos(\theta)\;\;-\sin(\theta)\,] + [\,Ax\;\;Ay\;\;Az\,] + [\,0\;-100\;\;0\,]$$
$$+ [\,0\;\;By\;\;Bz\,]$$

```
> evalm(%);
```

$$[\,Ax\;\;P\cos(\theta) + Ay - 100 + By\;\;-P\sin(\theta) + Az + Bz\,]$$

The summation of moments about the origin is zero.

```
> crossprod(vector([-1,sin(theta),cos(theta)]),
  P*vector([0,cos(theta),-sin(theta)]))+
  crossprod(vector([2,0,2/12]),vector([0,-100,0]))+
  crossprod(vector([4,0,0]),vector([0,By,Bz]));
```

Computational Solution — Sample Problem 6.7 — Symbolic Generation and Solution of Equilibrium Equations

$$\left[-\sin(\theta)^2 P - \cos(\theta)^2 P \; -P\sin(\theta) \; -P\cos(\theta) \right]$$

$$+ \left[\frac{50}{3} \quad 0 \quad -200 \right] + \left[0 \quad -4Bz \quad 4By \right]$$

The previous expression can be simplified using the *simplify* function.

```
> simplify(evalm(%));
```

$$\left[\frac{50}{3} - P \quad -P\sin(\theta) - 4Bz \quad -P\cos(\theta) - 200 + 4By \right]$$

Setting the scalar components of the two vector equations equal to zero and solving yields the following.

```
> eq1:=Ax=0:
> eq2:=P*cos(theta)+Ay-100+By=0:
> eq3:=-P*sin(theta)+Az+Bz=0:
> eq4:=-P+50/3=0:
> eq5:=-P*sin(theta)-4*Bz=0:
> eq6:=-P*cos(theta)-200+4*By=0:
> evalf(solve({eq1,eq2,eq3,eq4,eq5,eq6},
   {P,Ax,Ay,Az,By,Bz}));
```

$$\{Bz = -4.166666667\sin(\theta), P = 16.66666667, Ax = 0,$$
$$Az = 20.83333333\sin(\theta), By = 4.166666667\cos(\theta)$$
$$+ 50, Ay = -20.83333333\cos(\theta) + 50\}$$

The bearing forces can be graphed for a complete shaft revolution to determine when they attain their maximum values.

For a complete revolution:

```
> Ay:=theta->-125/6*cos(theta)+50:
> Az:=theta->20.82*sin(theta):
> plot1:=plot([[theta,Ay(theta*Pi/180),theta=0..400],
   [theta,Az(theta*Pi/180),theta=0..400]],x=0..400,
   y=-50..100,labels=[`theta(degrees)`,``],
   color=[black,black],linestyle=[SOLID,DOT]):
> textplot1:=textplot([200,80,"Ay(theta)"],
   labelfont=[TIMES,BOLD,12]):
> textplot2:=textplot([140,30,"Az(theta)"],
   labelfont=[TIMES,BOLD,12]):
> display([plot1,textplot1,textplot2]);
```

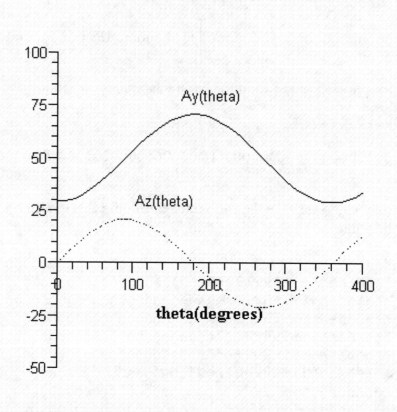

The magnitude of the bearing force at *A* is:

```
> A:=theta->
  sqrt(dotprod([0,Ay(theta),Az(theta)],
  [0,Ay(theta),Az(theta)])):
> plot([theta,A(theta*Pi/180),theta=0..400],x=0..400,
  y=20..80,labels=[`theta(degrees)`,`A(theta)`],
  labelfont=[TIMES,BOLD,12],color=black);
```

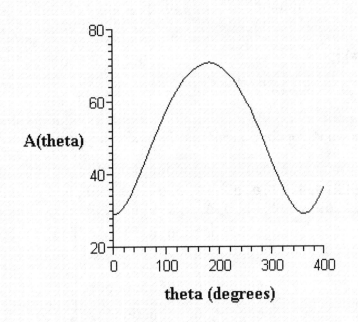

Computational Solution — Sample Problem 6.8 — Computational Generation of Equilibrium Equations

The resultant vector is equal to zero.

```
> restart;
> with(linalg):
> vector([0,-k*ys,0])+vector([Ax,Ay,Az])+vector([Bx,By,0]);
```

$$\begin{bmatrix} 0 & -k\,ys & 0 \end{bmatrix} + \begin{bmatrix} Ax & Ay & Az \end{bmatrix} + \begin{bmatrix} Bx & By & 0 \end{bmatrix}$$

```
> evalm(%);
```

$$\begin{bmatrix} Ax + Bx & -k\,ys + Ay + By & Az \end{bmatrix}$$

The summation of moments is equal to zero.

```
> crossprod(vector([xs,ys,0]),vector([0,-k*ys,0]))+
  crossprod(vector([0,0,1]),vector([Ax,Ay,Az]))+
  crossprod(vector([0,0,2]),vector([Bx,By,0]))+
  vector([0,0,T]);
```

$$\begin{bmatrix} 0 & 0 & -xs\,k\,ys \end{bmatrix} + \begin{bmatrix} -Ay & Ax & 0 \end{bmatrix} + \begin{bmatrix} 2\,By & -2\,Bx & 0 \end{bmatrix}$$
$$+ \begin{bmatrix} 0 & 0 & T \end{bmatrix}$$

> `evalm(%);`

$$\begin{bmatrix} -Ay + 2\,By & Ax - 2\,Bx & -xs\,k\,ys + T \end{bmatrix}$$

The six unknown vector components may be determined symbolically in terms of the spring force.

> `A:=matrix(6,6,[1,0,0,1,0,0,0,1,0,0,1,0,0,0,1,0,0,0,0,`
> `-1,0,0,2,0,1,0,0,-2,0,0,0,0,0,0,1]);`

$$A := \begin{bmatrix} 1 & 0 & 0 & 1 & 0 & 0 \\ 0 & 1 & 0 & 0 & 1 & 0 \\ 0 & 0 & 1 & 0 & 0 & 0 \\ 0 & -1 & 0 & 0 & 2 & 0 \\ 1 & 0 & 0 & -2 & 0 & 0 \\ 0 & 0 & 0 & 0 & 0 & 1 \end{bmatrix}$$

> `B:=matrix(6,1,[0,k*ys,0,0,0,k*xs*ys]);`

$$B := \begin{bmatrix} 0 \\ k\,ys \\ 0 \\ 0 \\ 0 \\ xs\,k\,ys \end{bmatrix}$$

> `[Ax,Ay,Az,Bx,By,T]=transpose(multiply(inverse(A),B));`

$$([Ax, Ay, Az, Bx, By, T]) = \begin{bmatrix} 0 & \frac{2}{3}\,k\,ys & 0 & 0 & \frac{1}{3}\,k\,ys & xs\,k\,ys \end{bmatrix}$$

Computational Solution — Sample Problem 6.8

The nonzero force components can be graphed for one full cam revolution.

```
> restart;
> with(plots):
> delta:=.2: r:=.4: k:=200:
> xs:=theta->delta*cos(theta):
> ys:=theta->r+delta*sin(theta):
> Ay:=theta->2/3*k*ys(theta):
> By:=theta->(k*ys(theta))/3:
> T:=theta->k*xs(theta)*ys(theta):
> plot1:=plot([[theta,Ay(theta*Pi/180),theta=0..400],
    [theta,By(theta*Pi/180),theta=0..400]],x=0..400,y=0..100,
    labels=[`theta(degrees)`,``],labelfont=[TIMES,BOLD,12],
    color=[black,black],linestyle=[SOLID,DOT]):
> textplot1:=textplot([200,70,"Ay(theta)"],
    labelfont=[TIMES,BOLD,12]):
> textplot2:=textplot([100,30,"By(theta)"],
    labelfont=[TIMES,BOLD,12]):
> display(plot1,textplot1,textplot2);
```

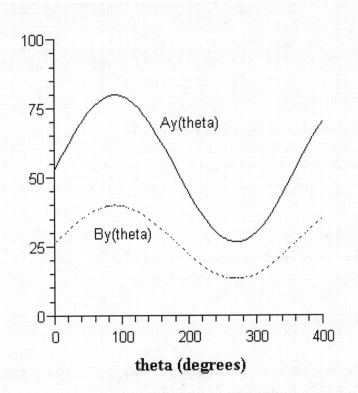

```
> plot([theta,T(theta*Pi/180),theta=0..400],x=0..400,
  y=-20..20,labels=[`theta(degrees)`,`T(theta)`],
  labelfont=[TIMES,BOLD,12],color=black);
```

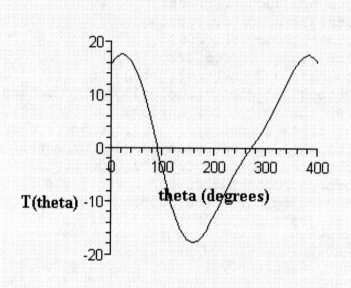

Computational Solution — Sample Problem 6.10

The three simultaneous equations in Sample Problem 6.10 may be solved symbolically using *Maple*.

```
> restart;
> with(linalg):
> matrix(3,3,[-1,2,-1,1,1,1,0,L,2*L]);
```

$$\begin{bmatrix} -1 & 2 & -1 \\ 1 & 1 & 1 \\ 0 & L & 2L \end{bmatrix}$$

```
> [Fa,Fb,Fc]=simplify(transpose(multiply(inverse(%),
  matrix(3,1,[0,P,alpha*P]))));
```

$$([Fa, Fb, Fc]) = \begin{bmatrix} \frac{1}{6}\frac{P(5L-3\alpha)}{L} & \frac{1}{3}P & -\frac{1}{6}\frac{P(L-3\alpha)}{L} \end{bmatrix}$$

Computational Solution – Sample Problem 6.11

The system of six equations is solved using matrix methods.

```
> restart;
> with(linalg):
> coef:=matrix(6,6,[1,1,1,1,0,0,0,6,6,0,0,0,0,0,3,3,0,0,
  1,-1,0,0,0,6,1,0,-1,0,3,6,1,0,0,-1,3,0]);
```

$$coef := \begin{bmatrix} 1 & 1 & 1 & 1 & 0 & 0 \\ 0 & 6 & 6 & 0 & 0 & 0 \\ 0 & 0 & 3 & 3 & 0 & 0 \\ 1 & -1 & 0 & 0 & 0 & 6 \\ 1 & 0 & -1 & 0 & 3 & 6 \\ 1 & 0 & 0 & -1 & 3 & 0 \end{bmatrix}$$

```
> [A,B,C,D,alpha,beta]=transpose(evalf(multiply(
  inverse(coef),matrix(6,1,[80,181,105,0,0,0]))));
```

$$([A,B,C,D,\alpha,\beta]) = [27.41666667, 17.58333333, 12.58333333$$
$$22.41666667, -1.666666667, -1.638888889]$$

Computational Solution — Sample Problem 6.12 — Symbolic Generation of Equilibrium Equations

In some cases, it may be useful to use the symbolic processor to generate equilibrium equations. The equations should be defined symbolically in vector form. This procedure does not offer many advantages in the force equation but may be very useful in the moment equation because the cross product of the unknown forces would be formed symbolically. An example of this approach is shown for Sample Problem 6.12.

The following expression is the force equilibrium equation.

```
> restart;
> with(linalg):
> vector([Ax,Ay,Az])+vector([Bx,By,Bz])+T*vector([
  -.156,.312,-.937])+vector([0,-981,0])=0;
```

$$\begin{bmatrix} Ax & Ay & Az \end{bmatrix} + \begin{bmatrix} Bx & By & Bz \end{bmatrix} + T\begin{bmatrix} -.156 & 0.312 & -.937 \end{bmatrix}$$
$$+ \begin{bmatrix} 0 & -981 & 0 \end{bmatrix} = 0$$

```
> evalm(%);
```

$$\left[[Ax + Bx - 0.156\,T, Ay + By + 0.312\,T - 981, Az + Bz - 0.937]\right] = 0$$

The following expression set equal to zero is the moment equilibrium equation.

```
> crossprod(vector([0,0,3]),vector([Ax,Ay,Az]))+
  crossprod(vector([3,0,0]),vector([Bx,By,Bz]))+
  crossprod(vector([2,0,3]),T*vector([-.156,.312,-.937]))+
  crossprod(vector([3,0,3]),vector([0,-981,0]))=0;
```

$$\begin{bmatrix} -3\,Ay & 3\,Ax & 0 \end{bmatrix} + \begin{bmatrix} 0 & -3\,Bz & 3\,By \end{bmatrix}$$
$$+ \begin{bmatrix} -0.936\,T & 1.406\,T & 0.624\,T \end{bmatrix} + \begin{bmatrix} 2943 & 0 & -2943 \end{bmatrix} = 0$$

```
> evalm(%);
```

$$\left[[-3\,Ay - 0.936\,T + 2943,\ 3\,Ax - 3\,Bz + 1.406\,T,\right.$$
$$\left. 3\,By + 0.624\,T - 2943]\right] = 0$$

As discussed in the text, this is a system of six equations for seven unknowns and the problem as modeled is improperly supported. Remodeling the structure yields a system of six equations for six unknowns. The new set of equilibrium equations can be generated symbolically when the reactions at A and B are represented in normal and perpendicular coordinates. The moment of the tension and the weight will not be recalculated but copied from the above symbolic representation.

The following expression set equal to zero is the force equilibrium equation.

```
> vector([.707*Ab+.707*Ap,Ay,-.707*Ab+.707*Ap])+
  vector([.707*Bp,By,.707*Bp])+
  T*vector([-.156,.312,-.937])+vector([0,-981,0])=0;
```

$$\begin{bmatrix} 0.707\,Ab + 0.707\,Ap & Ay & -0.707\,Ab + 0.707\,Ap \end{bmatrix}$$
$$+ \begin{bmatrix} 0.707\,Bp & By & 0.707\,Bp \end{bmatrix} + T\begin{bmatrix} -.156 & 0.312 & -.937 \end{bmatrix}$$
$$+ \begin{bmatrix} 0 & -981 & 0 \end{bmatrix} = 0$$

```
> evalm(%);
```

Computational Solution — Sample Problem 6.12 — Numerical Solution

$$\left[[0.707\, Ab + 0.707\, Ap + 0.707\, Bp - 0.156\, T, \right.$$
$$Ay + By + 0.312\, T - 981,$$
$$\left. -0.707\, Ab + 0.707\, Ap + 0.707\, Bp - 0.937\, T] \right] = 0$$

The following expression set equal to zero is the moment equilibrium equation.

```
> crossprod(vector([0,0,3]),vector([.707*Ab+.707*Ap,Ay,
  -.707*Ab+.707*Ap]))+crossprod(vector([3,0,0]),
  vector([.707*Bp,By,.707*Bp]))+T*vector([-
  .936,1.406,.624])+vector([2943,0,-2943])=0;
```

$$\left[-3\,Ay \quad 2.121\,Ab + 2.121\,Ap \quad 0 \right] + \left[0 \quad -2.121\,Bp \quad 3\,By \right]$$
$$+ T\left[-.936 \quad 1.406 \quad 0.624 \right] + \left[2943 \quad 0 \quad -2943 \right] = 0$$

```
> evalm(%);
```

$$\left[[-3\,Ay - 0.936\,T + 2943, \right.$$
$$2.121\,Ab + 2.121\,Ap - 2.121\,Bp + 1.406\,T,$$
$$\left. 3\,By + 0.624\,T - 2943] \right] = 0$$

This linear system of equations is solved by writing the six equations in matrix notation and using matrix methods. The matrix elements are entered from these equilibrium equations and the solution of the linear system is independent of the symbolic generation of these equations.

Computational Solution — Sample Problem 6.12 — Numerical Solution

```
> restart;
> with(linalg):
> C:=matrix(6,6,[.707,0,.707,0,.707,-.156,0,1,0,1,0,.312,
  -.707,0,.707,0,.707,-.937,0,-3,0,0,0,-.936,2.121,0,2.121,
  0,-2.121,1.406,0,0,0,3,0,.624]);
```

$$C := \begin{bmatrix} 0.707 & 0 & 0.707 & 0 & 0.707 & -.156 \\ 0 & 1 & 0 & 1 & 0 & 0.312 \\ -.707 & 0 & 0.707 & 0 & 0.707 & -.937 \\ 0 & -3 & 0 & 0 & 0 & -.936 \\ 2.121 & 0 & 2.121 & 0 & -2.121 & 1.406 \\ 0 & 0 & 0 & 3 & 0 & 0.624 \end{bmatrix}$$

```
> W:=matrix(6,1,[0,981,0,-2943,0,2943]);
```

$$W := \begin{bmatrix} 0 \\ 981 \\ 0 \\ -2943 \\ 0 \\ 2943 \end{bmatrix}$$

```
> [An,Ay,Ap,By,Bp,T]=transpose(multiply(inverse(C),W));
```

$([An, Ay, Ap, By, Bp, T]) = [-2604.997416, -490.5000001,$

$1562.108991, 1.000001089 \; 10^{-7}, 2083.553203, 4716.346156]$

Note that although the value of B_y appears to be nonzero, its value is extremely small and is nonzero only because of round-off error.

The tension in the cable could have been determined in both the statically determinate and the statically indeterminate cases by setting the component of the moment about the line AB equal to zero. The reactions at A and B do not produce moments about the line from A to B. This solution is used sometimes but is based upon a special observation and is not the general solution. The moment about the line AB must be zero for equilibrium and the tension is determined in the following solution.

Computational Solution — Sample Problem 6.12 — Numerical Solution for the Tension in the Cable

This solution is not a full solution and is obtained by setting the component of the moment about the line AB equal to zero.

```
> restart;
> with(linalg):
> AB:=vector([3,0,-3]):
> eAB:=normalize(AB):
```

Computational Solution — Sample Problem 6.12 — Numerical Solution for the Tension in the Cable

Taking the moment of **T** and **W** about point A and taking the scalar product with the unit vector along AB yields:

```
> dotprod((crossprod(vector([2,0,0]),
  T*vector([-.156,.312,-.937])))+
  crossprod(vector([3,0,0]),vector([0,-981,0]))),eAB)=0;
```

$$-\frac{1}{2}(0.624\,T - 2943)\sqrt{2} = 0$$

Setting the previous expression equal to zero and solving yields:

```
> T:=solve(%,T);
```

$$T := 4716.346154$$

The tension in the cable is the same as previously determined.

7 Analysis of Structures

Application of equilibrium principles to structures is illustrated in this chapter. The structures include plane and space trusses, frames, and machines. Most modern texts in structural analysis present methods using a matrix approach. The simplest example of this approach is discussed in Section 7.5 of the text. A symbolic solution for the plane three-member truss in Figure 7.18 of the text is shown in Computational Window 7.1. The symbolic solution can be generated using *Maple*.

Computational Window 7.1

The matrix characterizing the three-member truss is entered for any angles α and β, inverted, and then multiplied by the negative of the joint loads.

```
> restart;
> with(linalg):
> coef:=matrix(6,6,[cos(alpha),1,0,1,0,0,sin(alpha),0,0,0,
  1,0,-cos(alpha),0,cos(beta),0,0,0,-sin(alpha),0,
  -sin(beta),0,0,0,0,-1,-cos(beta),0,0,0,0,0,sin(beta),
  0,0,1]):
> const:=matrix(6,1,[-pax,-pay,-pbx,-pby,-pcx,-pcy]):
```

The equation is evaluated symbolically as shown. (The six values are displayed individually to simplify the presentation.)

```
> simplify(multiply(inverse(coef),const))[1,1];
```
$$\frac{\sin(\beta)\,pbx + \cos(\beta)\,pby}{\cos(\alpha)\sin(\beta) + \sin(\alpha)\cos(\beta)}$$

```
> simplify(multiply(inverse(coef),const))[2,1];
```
$$\frac{1}{\cos(\alpha)\sin(\beta) + \sin(\alpha)\cos(\beta)}(\sin(\alpha)\cos(\beta)\,pbx$$
$$- \cos(\alpha)\cos(\beta)\,pby + pcx\cos(\alpha)\sin(\beta)$$
$$+ pcx\sin(\alpha)\cos(\beta))$$

```
> simplify(multiply(inverse(coef),const))[3,1];
```
$$-\frac{\sin(\alpha)\,pbx - \cos(\alpha)\,pby}{\cos(\alpha)\sin(\beta) + \sin(\alpha)\cos(\beta)}$$

```
> simplify(multiply(inverse(coef),const))[4,1];
```
$$-pax - pbx - pcx$$

Computational Window 7.2

```
> simplify(multiply(inverse(coef),const))[5,1];
```

$$-\frac{1}{\cos(\alpha)\sin(\beta)+\sin(\alpha)\cos(\beta)}(pay\cos(\alpha)\sin(\beta)$$
$$+pay\sin(\alpha)\cos(\beta)+\sin(\alpha)\sin(\beta)pbx$$
$$+\sin(\alpha)\cos(\beta)pby)$$

```
> simplify(multiply(inverse(coef),const))[6,1];
```

$$\frac{1}{\cos(\alpha)\sin(\beta)+\sin(\alpha)\cos(\beta)}(\sin(\alpha)\sin(\beta)pbx$$
$$-\cos(\alpha)\sin(\beta)pby-pcy\cos(\alpha)\sin(\beta)-pcy\sin(\alpha)\cos(\beta))$$

The truss shown in Figure 7.18 is examined for particular loadings in Computational Windows 7.2 and 7.3. In the first case, a unit vertical load P_{By} is applied and the internal forces in AB and AC and the vertical reaction A_y as a function of the angle α are determined when $\alpha = \beta$. In the second case, the load is applied at B at an angle of 30 degrees with the vertical and the internal forces in AB, AC, and BC are examined as a function of the angle α when $\alpha = \beta$.

Computational Window 7.2

Consider the case when the triangle is symmetric. The two angles are equal in this case and are varied in the solution from 5 to 85 degrees. (The truss will be unstable at 0 and 90 degrees.) A unit positive vertical load is applied at B.

```
> restart;
> with(plots):
> AB:=alpha->cos(alpha)/(2*sin(alpha)*cos(alpha)):
> AC:=alpha->-(cos(alpha))^2/(2*sin(alpha)*cos(alpha)):
> Ay:=alpha->(-sin(alpha)*cos(alpha))/
  (2*sin(alpha)*cos(alpha)):
```

The two members AB and BC will have equal internal forces in this symmetric case and the vertical reactions at A and C will also be equal.

```
> plot1:=plot([alpha,AB(alpha*Pi/180),alpha=0..100],
  x=0..100,y=-10..10,color=black,linestyle=DOT):
> plot2:=plot([alpha,AC(alpha*Pi/180),alpha=0..100],
  x=0..100,y=-10..10,color=black,linestyle=DASH):
> plot3:=plot([alpha,Ay(alpha*Pi/180),alpha=0..100],
  x=0..100,y=-10..10,color=black,linestyle=SOLID):
```

```
> textplot1:=textplot([16,6,"AB(alpha)"],
  labelfont=[TIMES,BOLD,12]):
> textplot2:=textplot([14,-9,"AC(alpha)"],
  labelfont=[TIMES,BOLD,12]):
> textplot3:=textplot([87,-2,"Ay(alpha)"],
  labelfont=[TIMES,BOLD,12]):
> display(plot1,plot2,plot3,textplot1,textplot2,textplot3,
  labels=[`alpha(degrees)`,``]);
```

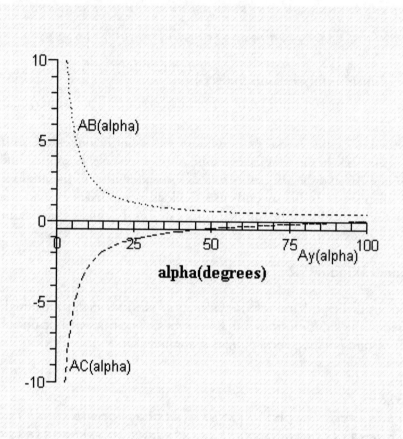

```
> AB45:=AB(45*Pi/180); AC45:=AC(45*Pi/180);
  Ay45:=Ay(45*Pi/180);
```

$$AB45 := \frac{1}{2}\sqrt{2}$$

$$AC45 := -\frac{1}{2}$$

$$Ay45 := -\frac{1}{2}$$

Computational Window 7.3

Consider the case when the triangle is symmetric. The two angles are equal in this case and are varied in the solution from 5 to 85 degrees. The truss will be unstable at 0 and 90 degrees. A unit load is applied at *B* at an angle of 30 degrees with the vertical.

```
> restart;
> with(plots):
> PBx:=.5: PBy:=.866:
> AB:=alpha->
  (PBx*sin(alpha)+PBy*cos(alpha))/(2*sin(alpha)*cos(alpha)):
> AC:=alpha->(PBx*sin(alpha)*cos(alpha)-
  PBy*(cos(alpha))^2)/(2*sin(alpha)*cos(alpha)):
> BC:=alpha->(-PBx*sin(alpha)+PBy*cos(alpha))/
  (2*sin(alpha)*cos(alpha)):
```

The internal forces of the members as a function of the truss angle are shown to be the following.

```
> plot1:=plot([alpha,AB(alpha*Pi/180),alpha=0..90],x=0..90,
  y=-6..6,color=black,linestyle=DOT):
> plot2:=plot([alpha,AC(alpha*Pi/180),alpha=0..90],x=0..90,
  y=-6..6,color=black,linestyle=DASH):
> plot3:=plot([alpha,BC(alpha*Pi/180),alpha=0..90],x=0..90,
  y=-6..6,color=black,linestyle=SOLID):
> textplot1:=textplot([20,3,"AB(alpha)"],
  labelfont=[TIMES,BOLD,12]):
> textplot2:=textplot([20,-2,"AC(alpha)"],
  labelfont=[TIMES,BOLD,12]):
> textplot3:=textplot([78,-4,"BC(alpha)"],
  labelfont=[TIMES,BOLD,12]):
> display(plot1,plot2,plot3,textplot1,textplot2,textplot3,
  labels=[`alpha (degrees)`,``],labelfont=[TIMES,BOLD,12]);
```

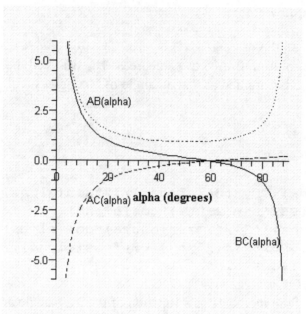

The graph shows the internal forces of the members as a function of the truss angle. Modern software has outdated many of the old methods involving techniques that were developed primarily to reduce computational difficulties. For example, the method of sections presented in the text is seldom used in structural analysis. However, *Maple* can be used to reduce the burden of solving the system of linear equations that arise when the method of sections is used. The system of three equations generated from Figure 7.21 is solved using linear algebra in the plane truss problem.

Computational Solution — Figure 7.21 — Method of Sections

```
> restart;
> with(linalg):
> Ay:=14.32:
> coefficient_matrix:=matrix(3,3,[1,3/5,cos(45*Pi/180),0,
  -4/5,sin(45*Pi/180),4,0,0]):
> constant_matrix:=matrix(3,1,[0,10-Ay,4*Ay-10]):
> [CF,DF,DE]=transpose(evalf(multiply(inverse(
  coefficient_matrix),constant_matrix)));
```

$$([CF, DF, DE]) = [11.82000000, -5.357142857, -12.17031785$$

Computational Solution — Sample Problem 7.2 — Case 1

Sample Problem 7.2 requires the solution of a system of 12 linear equations and is solved using the *fsolve* function. This system could be solved using matrix methods, but since the coefficient matrix is large and sparse, it requires somewhat less input if the *Maple* function for solving systems of equations is used. The system of equations is solved for the loading shown in the text and for the case when an additional 2000 N load in the

Computational Solution — Sample Problem 7.2 — Case 2

x-direction is placed at joint *B*. In the second case, the fourth equation would change. This equation can be changed and the solution easily obtained.

```
> restart;
> eq1:=-.447*AB-AC-.8*AD=0:
> eq2:=.6*AD+Ay=0:
> eq3:=.894*AB+Az=0:
> eq4:=.447*AB=0:
> eq5:=.352*BD+2000=0:
> eq6:=-.894*AB-CB-.936*BD=0:
> eq7:=AC+Cx=0:
> eq8:=CD+Cy=0:
> eq9:=CB+Cz=0:
> eq10:=.8*AD=0:
> eq11:=-.6*AD-.351*BD-CD=0:
> eq12:=.936*BD+Dz=0:
> fsolve({eq1,eq2,eq3,eq4,eq5,eq6,eq7,eq8,eq9,eq10,eq11,
  eq12},{AB,AC,AD,CB,BD,CD,Ay,Az,Cx,Cy,Cz,Dz});
```

$$\{AB = -0, AD = -0, Cx = -0, BD = -5681.818182,$$
$$Cz = -5318.181818, Cy = -1994.318182, Dz = 5318.181818,$$
$$CB = 5318.181818, CD = 1994.318182, AC = 0, Az = 0,$$
$$Ay = 0\}$$

Computational Solution — Sample Problem 7.2 — Case 2

```
> restart;
> eq1:=-.447*AB-AC-.8*AD=0:
> eq2:=.6*AD+Ay=0:
> eq3:=.894*AB+Az=0:
> eq4:=.447*AB+2000=0:
> eq5:=.352*BD+2000=0:
> eq6:=-.894*AB-CB-.936*BD=0:
> eq7:=AC+Cx=0:
> eq8:=CD+Cy=0:
> eq9:=CB+Cz=0:
> eq10:=.8*AD=0:
> eq11:=-.6*AD-.351*BD-CD=0:
> eq12:=.936*BD+Dz=0:
> fsolve({eq1,eq2,eq3,eq4,eq5,eq6,eq7,eq8,eq9,eq10,eq11,
  eq12},{AB,AC,AD,CB,BD,CD,Ay,Az,Cx,Cy,Cz,Dz});
```

$$\{AD = -0, AB = -4474.272930, BD = -5681.818182,$$
$$Cz = -9318.181817, Cx = -2000.000000, Cy = -1994.318182$$
$$Dz = 5318.181818, CB = 9318.181817, CD = 1994.318182,$$
$$AC = 2000.000000, Az = 3999.999999, Ay = 0\}$$

Computational Solution — Sample Problem 7.3 — Motor Torque Graph ($L_2 = 200$ mm)

To fully understand the solution of Sample Problem 7.3, the motor torque should be plotted for a full revolution. In addition, the solution dependence on the length L_1 and L_2 can be investigated.

The required torque for equilibrium for the numerical values given is easily determined using *Maple*. Note that L_1 must be greater than L_2 or the machine will bind before rotating 90 degrees. While member L_2 rotates through 360 degrees, member L_1 oscillates through a fixed arc.

```
> restart;
> L1:=500: L2:=200: P:=1000:
> theta:=beta->arcsin((L2/L1)*sin(beta)):
> Mo:=beta->
  (L1*cos(theta(beta))+L2*cos(beta))*P*tan(theta(beta)):
> plot([beta,Mo(beta*Pi/180),beta=0..360],x=0..7*180/Pi,
  y=-4*10^5..4*10^5,labels=[`beta(degrees)`,`Mo(beta)`],
  color=black,labelfont=[TIMES,BOLD,12]);
```

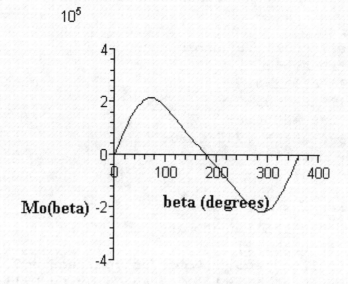

Note that the moment required for equilibrium is zero when the piston is at the top or bottom of its stroke and that the moment is negative during the second half of the cycle.

Computational Solution — Sample Problem 7.3 — Motor Torque Graph ($L_2 = 300$ mm)

The effect of changing the ratio of L_1 to L_2 can be easily examined. Once the *Maple* problem statement is created, a change in a constant such as L_2 can be accomplished simply by changing the value in the declaration. Suppose the length of L_2 is changed to 300 mm, the required moment would increase as shown.

```
> restart;
> L1:=500: L2:=300: P:=1000:
> theta:=beta->arcsin((L2/L1)*sin(beta)):
> Mo:=beta->
  (L1*cos(theta(beta))+L2*cos(beta))*P*tan(theta(beta)):
> plot([beta,Mo(beta*Pi/180),beta=0..360],x=0..7*180/Pi,
  y=-4*10^5..4*10^5,labels=[`beta(degrees)`,`Mo(beta)`],
  color=black,labelfont=[TIMES,BOLD,12]);
```

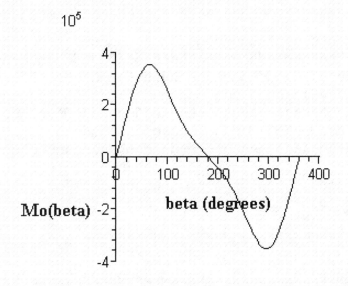

Again, note that the moment required for equilibrium is zero when the piston is at the top or bottom of its stroke and that the moment is negative during the second half of the cycle. It is clear that the moment required for equilibrium increases as the length of L_2 increases.

Sample Problem 7.4 is solved using different combinations of the equilibrium equations.

Computational Solution — Sample Problem 7.4 — Equations (a) and (b)

Solving symbolically using equations (a) and (b) yields the following.

```
> restart;
> with(linalg):
> coefficient_matrix:=matrix(6,6,[1,0,1,0,0,0,0,1,0,1,0,-
  1,0,0,25,25,0,-55,0,0,-1,0,0,0,0,0,0,-1,1,1,0,0,0,0,-
  35,30]):
> constant_matrix:=matrix(6,1,[0,F,-85*F,0,0,0]):
> [Cx,Cy,Dx,Dy,BH,P]=transpose(multiply(inverse(
  coefficient_matrix),constant_matrix));
```

$$([Cx, Cy, Dx, Dy, BH, P]) = [0, -\frac{15}{2}F, 0, \frac{221}{12}F, \frac{17}{2}F, \frac{119}{12}F]$$

Computational Solution — Sample Problem 7.4 — Equations (a) and (c)

Solving symbolically using equations (a) and (c) yields the following.

```
> restart;
> with(linalg):
> coefficient_matrix:=matrix(6,6,[1,0,1,0,0,0,0,1,0,1,0,-
  1,0,0,25,25,0,-55,1,0,0,0,0,0,0,-1,0,0,-1,0,0,-
  10,0,0,0,0]):
> constant_matrix:=matrix(6,1,[0,F,-85*F,0,-F,75*F]):
> [Cx,Cy,Dx,Dy,BH,P]=transpose(multiply(
  inverse(coefficient_matrix),constant_matrix));
```

$$([Cx, Cy, Dx, Dy, BH, P]) = [0, -\frac{15}{2}F, 0, \frac{221}{12}F, \frac{17}{2}F, \frac{119}{12}F]$$

Computational Solution — Sample Problem 7.4 — Equations (b) and (c)

Solving symbolically using equations (b) and (c) yields the following.

```
> restart;
> with(linalg):
```

Computational Solution — Sample Problem 7.4 — Equations (b) and (c)

```
> coefficient_matrix:=matrix(6,6,[0,0,-1,0,0,0,0,0,0,-
  1,1,1,0,0,25,0,-35,30,1,0,0,0,0,0,0,-1,0,0,-1,0,0,-
  10,0,0,0,0]):
> constant_matrix:=matrix(6,1,[0,0,0,0,-F,75*F]):
> [Cx,Cy,Dx,Dy,BH,P]=transpose(multiply(
  inverse(coefficient_matrix),constant_matrix));
```

$$([Cx, Cy, Dx, Dy, BH, P]) = [0, -\frac{15}{2}F, 0, \frac{221}{12}F, \frac{17}{2}F, \frac{119}{12}F]$$

8 Internal Forces in Structural Members

The principle of equilibrium is used in Chapter 8 to determine and diagram the internal forces in structural members. This chapter presents the basic concepts of beam loading in preparation for Mechanics of Materials courses. Most of the integrals required to determine the internal shear forces and moments in beams can be performed numerically or symbolically using *Maple*. The shear and moment diagrams or plots can then be generated. Sample Problem 8.4 is solved using both methods.

Computational Solution — Sample Problem 8.4 — Symbolic Solution

The integrals are evaluated symbolically but can be evaluated numerically and the shear and moment diagrams can be generated numerically.

```
> restart;
> Va:=int((W/L)*u,u=0..x);
```

$$Va := \frac{1}{2} \frac{W x^2}{L}$$

```
> Ma:=int((W/L)*(x-u)*u,u=0..x);
```

$$Ma := \frac{1}{6} \frac{W x^3}{L}$$

```
> Vb:=int(W*sin(Pi*u/L),u=0..x);
```

$$Vb := -\frac{W L \left(-1 + \cos\left(\frac{\pi x}{L}\right)\right)}{\pi}$$

```
> Mb:=int((x-u)*W*sin(Pi*u/L),u=0..x);
```

$$Mb := -\frac{W L \left(-\pi x + \sin\left(\frac{\pi x}{L}\right) L\right)}{\pi^2}$$

Computational Solution — Sample Problem 8.4 — Numerical Solution (Linear Loading)

```
> restart;
> W:=200: L:=10:
> w:=x->W*x/L:
> V:=x->int(w(u),u=0..x):
> M:=x->int((x-u)*w(u),u=0..x):
> plot([x,w(x),x=0..10],x=0..10,y=0..200,
  labels=[`x`,`w(x)`],labelfont=[TIMES,BOLD,12],
  color=black);
```

Computational Solution — Sample Problem 8.4 — Numerical Solution (Linear Loading)

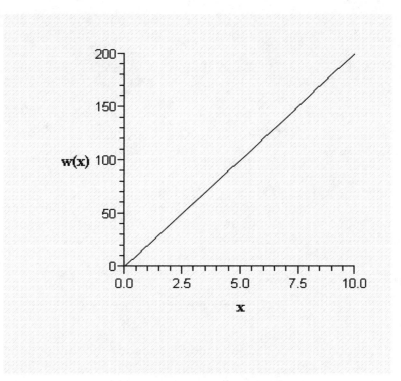

```
> plot([x,V(x),x=0..10],x=0..10,y=0..1000,
  labels=[`x`,`V(x)`],labelfont=[TIMES,BOLD,12],
  color=black);
```

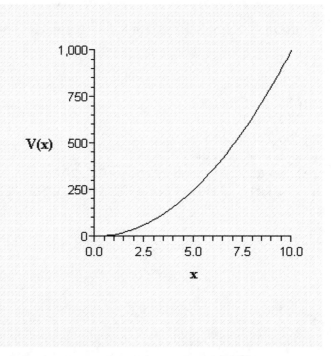

```
> plot([x,M(x),x=0..10],x=0..10,y=0..4000,
  labels=[`x`,`M(x)`],labelfont=[TIMES,BOLD,12],
  color=black);
```

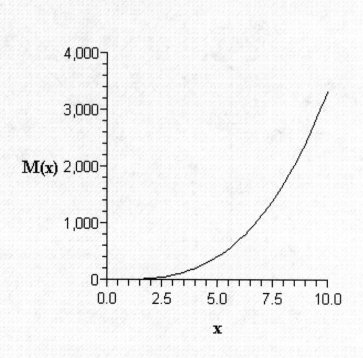

Computational Solution — Sample Problem 8.4 — Numerical Solution (Sinusoidal Loading)

```
> restart;
> W:=200: L:=10:
> w:=x->W*sin(Pi*x/L):
> V:=x->int(w(u),u=0..x):
> M:=x->int((x-u)*w(u),u=0..x):
> plot([x,w(x),x=0..10],x=0..10,y=0..200,
   labels=[`x`,`w(x)`],labelfont=[TIMES,BOLD,12],
   color=black);
```

Computational Solution — Sample Problem 8.4 — Numerical Solution (Sinusoidal Loading)   87

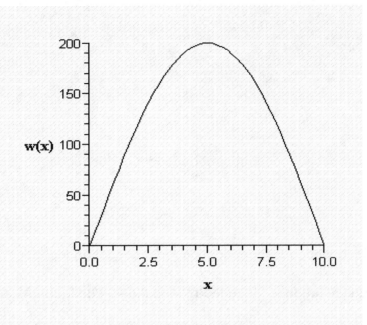

```
> plot([x,V(x),x=0..10],x=0..10,y=0..2000,
   labels=[`x`,`V(x)`],labelfont=[TIMES,BOLD,12],
   color=black);
```

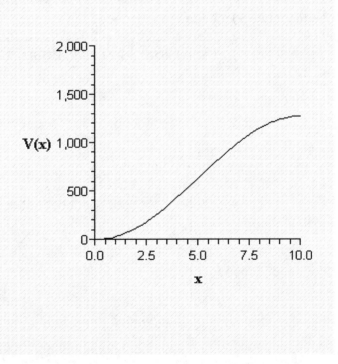

```
> plot([x,M(x),x=0..10],x=0..10,y=0..10000,
   labels=[`x`,`M(x)`],labelfont=[TIMES,BOLD,12],
   color=black);
```

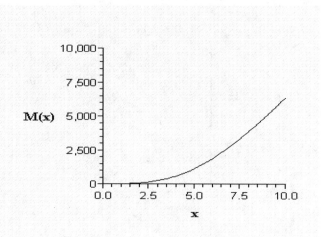

Computational Solution — Sample Problem 8.6 — Maximum Moment

The maximum moment in Sample Problem 8.6 can be determined using the *fsolve* function to solve the quadratic equation.

Graph the shear from $x = 5$ to $x = 10$ to obtain an approximation of the root.

```
> restart;
> V:=x->833-200*(x-5)^2/2:
> plot([x,V(x),x=4..10],x=4..10,y=-2000..1000,
  labels=[`x`,`V(x)`],labelfont=[TIMES,BOLD,12],
  color=black);
```

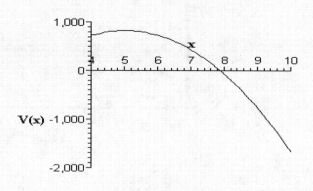

Look for the root between $x = 6$ and $x = 9$.

Computational Solution — Sample Problem 8.6 — Maximum Moment

```
> a:=fsolve(V(x),x,x=6..9);
```
$$a := 7.886173938$$

Solve for the value of the moment when $x = 7.88617$.

```
> M:=x->833*x-200*(x-5)^3/6:
> M(a):=M(a);
```
$$M(7.886173938) := 5767.788593$$

A graph of the moment equation for this region can also be created to facilitate drawing the moment diagram.

```
> plot([x,M(x),x=5..10],x=5..10,y=4000..6000,
  labels=[`x`,`M(x)`],labelfont=[TIMES,BOLD,12],
  color=black);
```

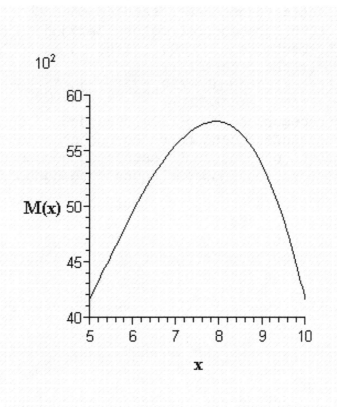

Computational Solution — Sample Problem 8.6 — Shear and Moment Diagrams

The shear and moment diagrams for Sample Problem 8.6 can be generated by defining each section of the beam as a separate plot. *Maple* allows the three section plots to be combined into one graph.

```
> restart;
> with(plots):
> V1:=x->833:
> M1:=x->833*x:
> V2:=y->833-200*(y-5)^2/2:
> M2:=y->833*y-200*(y-5)^3/6:
> V3:=z->-1667:
> M3:=z->-1667*z+20833:
> plot1:=plot([x,V1(x),x=0..5],x=0..14,
  y=-2000..1000,linestyle=1,color=black):
> plot2:=plot([y,V2(y),y=5..10],x=0..14,
  y=-2000..1000,linestyle=2,color=black):
> plot3:=plot([z,V3(z),z=10..12.5],x=0..14,
  y=-2000..1000,linestyle=1,color=black):
> textplot1:=textplot([3,1000,"V1(x)"],
  labelfont=[TIMES,BOLD,12]):
> textplot2:=textplot([8,-750,"V2(y)"],
  labelfont=[TIMES,BOLD,12]):
> textplot3:=textplot([11,-1500,"V3(z)"],
  labelfont=[TIMES,BOLD,12]):
> display(plot1,plot2,plot3,textplot1,textplot2,textplot3,
  labels=[`x, y, z`,``],labelfont=[TIMES,BOLD,12]);
```

Computational Solution — Sample Problem 8.6 — Shear and Moment Diagrams

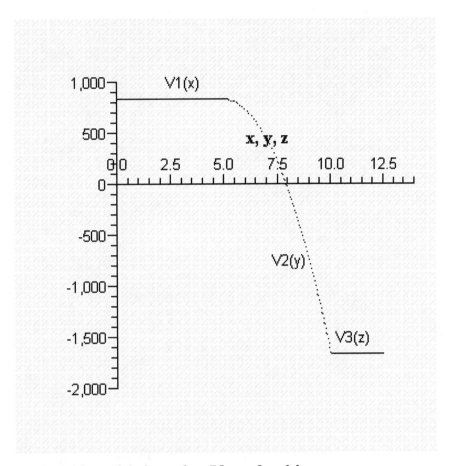

```
> plot1:=plot([x,M1(x),x=0..5],x=0..14,
  y=-2000..6000,linestyle=1,color=black):
> plot2:=plot([y,M2(y),y=5..10],x=0..14,
  y=-2000..6000,linestyle=2,color=black):
> plot3:=plot([z,M3(z),z=10..12.5],x=0..14,
  y=-2000..6000,linestyle=1,color=black):
> textplot1:=textplot([2,3000,"M1(x)"],
  labelfont=[TIMES,BOLD,12]):
> textplot2:=textplot([6,6000,"M2(y)"],
  labelfont=[TIMES,BOLD,12]):
> textplot3:=textplot([12,2500,"M3(z)"],
  labelfont=[TIMES,BOLD,12]):
> display(plot1,plot2,plot3,textplot1,textplot2,textplot3,
  labels=[`x, y, z`,``],labelfont=[TIMES,BOLD,12]);
```

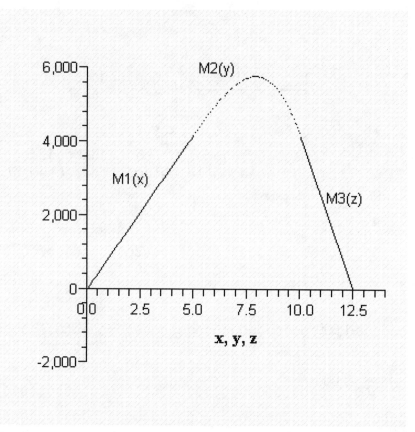

The maximum moment is determined by evaluating the moment equation at the root of the shear equation.

```
> V2root:=fsolve(V2(y),y,y=7.5..8.5);
```
$$V2root := 7.886173938$$

```
> M2ofV2root:=M2(V2root);
```
$$M2ofV2root := 5767.788593$$

Discontinuity Functions

Maple includes a definition for the Heaviside step function and it is designated by *Heaviside(t)*. For t less than zero the *Heaviside* function is equal to zero, for t equal to zero the *Heaviside* function is undefined, and for t greater than zero the *Heaviside* function is equal to one. Note that any algebraic expression may be used in the place of t. This allows the step point for the *Heaviside* function to be defined as any real number. For example, if the *Heaviside* function is to equal one for $t > 2$, it is necessary to use *Heaviside(t-2)*. In this case the argument of the *Heaviside* function is positive when $t > 2$ so the *Heaviside* function equals one as desired. *Maple* also includes a definition for the Dirac delta function and it is designated by *Dirac(t)*. The *Dirac* function has a value of zero at all t except at $t = 0$ at which it has a singularity. A second property of the *Dirac* function is that the integral of it from negative infinity to positive infinity is one. Or, in other words, the area under the *Dirac* function between negative infinity and positive infinity is one. The *Dirac* function is also the derivative of the *Heaviside* function.

Discontinuity Functions

Computational Solution — Sample Problem 8.7 — Shear and Moment Diagrams

```
> restart;
> V:=x->833*Heaviside(x)-100*(x-5)^2*Heaviside(x-5)+
  1000*(x-10)*Heaviside(x-10)+100*(x-10)^2*Heaviside(x-10):
```

Note that since the *Heaviside* function is undefined when $x = 0$, the range over which the shear function is evaluated must not include zero. However, a minimum x value may be chosen sufficiently close to zero so that no information is lost.

```
> plot([x,V(x),x=1*10^(-100)..12.5],x=0..14,y=-2000..1000,
  labels=[`x`,`V(x)`],labelfont=[TIMES,BOLD,12],
  color=black);
```

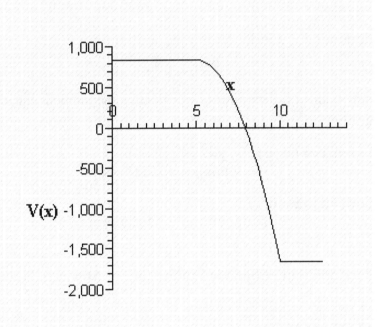

```
> M:=x->evalf(int(V(u),u=0..x)):
> plot([x,M(x),x=1*10^(-100)..12.5],x=0..14,
  y=-5000..10000,labels=[`x`,`M(x)`],
  labelfont=[TIMES,BOLD,12],color=black);
```

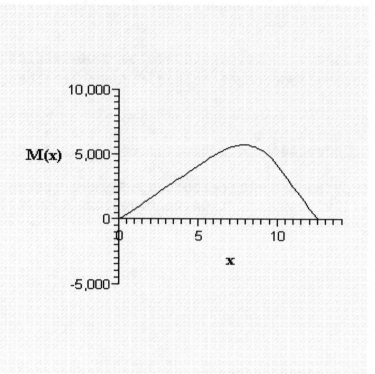

The maximum moment is determined by evaluating the moment equation at the root of the shear equation.

```
> Vroot:=fsolve(V(y),y,y=7.5..8.5);
```
$$Vroot := 7.886173938$$

```
> MofVroot:=M(Vroot);
```
$$MofVroot := 5767.788594$$

Computational Solution — Sample Problem 8.7 — Alternate Method

```
> restart;
> Rl:=833: Rr:=1667:
> V:=x->Rl*Heaviside(x)-Heaviside(x-5)*100*(x-5)^2
  +Heaviside(x-10)*100*(x-5)^2-2500*Heaviside(x-10):
> plot([x,V(x),x=1*10^(-100)..12.5],x=0..14,y=-2000..1000,
  labels=[`x`,`V(x)`],labelfont=[TIMES,BOLD,12],
  color=black);
```

Discontinuity Functions

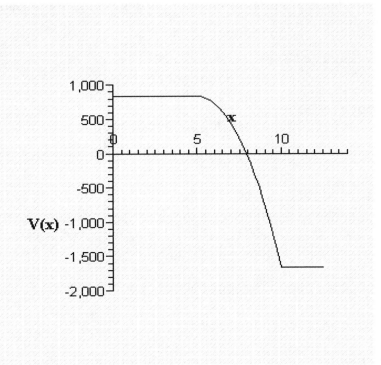

```
> M:=x->R1*x-Heaviside(x-5)*33.3*(x-5)^3+Heaviside(x-10)
  *33.3*(x-5)^3-Heaviside(x-10)*33.3*5^3-2500*(x-10)
  *Heaviside(x-10):
> plot([x,M(x),x=1*10^(-100)..12.5],x=0..14,y=-5000..10000,
  labels=[`x`,`M(x)`],labelfont=[TIMES,BOLD,12],
  color=black);
```

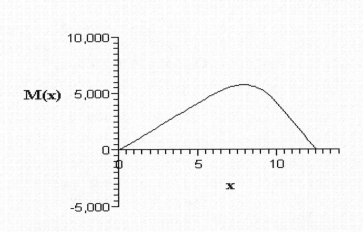

The maximum moment is determined by evaluating the moment equation at the root of the shear equation.

```
> Vroot:=fsolve(V(y),y,y=7.5..8.5);
```
$$Vroot := 7.886173938$$

```
> MofVroot:=M(Vroot);
```
$$MofVroot := 5768.589988$$

Cables

Problems involving cables loaded by concentrated forces result in systems of simultaneous nonlinear equations, as shown in Section 8.6 of the *Statics* text. Solution of these equations in *Maple* is accomplished by using the *fsolve* function, as shown in the Computational Solution of Sample Problem 8.8.

Computational Solution — Sample Problem 8.8 — L = 33 ft (Cable Length)

Note that angles are solved for in degrees.

```
> restart;
> P1:=750: P2:=400: ADy:=0: L:=33: ABx:=10: BCx:=10:
  CDx:=10:
> eq1:=-T1*cos(alpha1*Pi/180)+T2*cos(alpha2*Pi/180)=0:
> eq2:=-T1*sin(alpha1*Pi/180)+T2*sin(alpha2*Pi/180)-P1=0:
> eq3:=-T2*cos(alpha2*Pi/180)+T3*cos(alpha3*Pi/180)=0:
> eq4:=-T2*sin(alpha2*Pi/180)+T3*sin(alpha3*Pi/180)-P2=0:
> eq5:=ABx*tan(alpha1*Pi/180)+BCx*tan(alpha2*Pi/180)+
  CDx*tan(alpha3*Pi/180)=ADy:
> eq6:=ABx/cos(alpha1*Pi/180)+BCx/cos(alpha2*Pi/180)+
  CDx/cos(alpha3*Pi/180)=L:
> fsolve({eq1,eq2,eq3,eq4,eq5,eq6},{T1,T2,T3,alpha1,alpha2,
  alpha3},{T1=0..2000,T2=0..2000,T3=0..2000,alpha1=-
  90..90,alpha2=-90..90,alpha3=-90..90});
```
$$\{T1 = 1207.685424,\ T2 = 1034.893271,\ T3 = 1150.798599,$$
$$\alpha 1 = -31.62914727,\ \alpha 2 = 6.472887588,\ \alpha 3 = 26.67720817\}$$

After the solution has been set up in *Maple*, you can investigate the effects of varying parameters like the length of the cable. To change the cable length to 31 ft, change the length assignment statement. Re-evaluating each command using the keystroke [enter] will automatically show the effect of the change.

Cables   97

Computational Solution — Sample Problem 8.8 — $L = 31$ ft (Cable Length)

Note that angles are solved for in degrees.

```
> restart;
> P1:=750: P2:=400: ADy:=0: L:=31: ABx:=10: BCx:=10:
  CDx:=10:
> eq1:=-T1*cos(alpha1*Pi/180)+T2*cos(alpha2*Pi/180)=0:
> eq2:=-T1*sin(alpha1*Pi/180)+T2*sin(alpha2*Pi/180)-P1=0:
> eq3:=-T2*cos(alpha2*Pi/180)+T3*cos(alpha3*Pi/180)=0:
> eq4:=-T2*sin(alpha2*Pi/180)+T3*sin(alpha3*Pi/180)-P2=0:
> eq5:=ABx*tan(alpha1*Pi/180)+BCx*tan(alpha2*Pi/180)+
  CDx*tan(alpha3*Pi/180)=ADy:
> eq6:=ABx/cos(alpha1*Pi/180)+BCx/cos(alpha2*Pi/180)+
  CDx/cos(alpha3*Pi/180)=L:
> fsolve({eq1,eq2,eq3,eq4,eq5,eq6},{T1,T2,T3,alpha1,
  alpha2,alpha3},{T1=0..2000,T2=0..2000,T3=0..2000,
  alpha1=-90..90,alpha2=-90..90,alpha3=-90..90});
```

$$\{T1 = 1930.422086,\ T2 = 1827.301133,\ T3 = 1895.352939,$$
$$\alpha 1 = -19.15229497,\ \alpha 2 = 3.660621261,\ \alpha 3 = 15.81883513\}$$

Computational Solution — Sample Problem 8.9

The *fsolve* function can be used to solve Sample Problem 8.9. Establish the origin of the coordinate system at the lowest point on the cable and determine x_L (distance to the left support) and x_R (distance to the right support) from the equation of a parabola.

```
> restart;
> yL:=10: yR:=12: span:=50: w:=1000:
> eq1:=xL+xR-span=0:
> eq2:=yL-q*xL^2/2=0:
> eq3:=yR-q*xR^2/2=0:
> eq4:=To-w/q=0:
> fsolve({eq1,eq2,eq3,eq4},{xL,xR,q,To});
```

$$\{xR = 26.13872125,\ To = 28468.03119,\ q = 0.03512712184,$$
$$xL = 23.86127875\}$$

```
> To:=28468.03119:  q:=.03512712184:  xR:=26.13872125:
> T:=x->To*sqrt(1+q^2*x^2):
> TofxR:=T(xR);
```

$$TofxR := 38647.91777$$

```
> plot([x,T(x),x=-24..27],x=-40..40,y=2.5*10^4..4*10^4,
  labels=[`x`,`T(x)`],labelfont=[TIMES,BOLD,12],
  color=black);
```

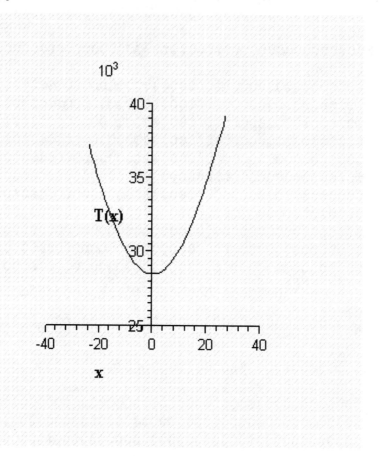

Computational Solution — Sample Problem 8.10

The catenary cable of length 20 m spans a distance of 18 m and has a weight of 40 N/m. The value of q is found by solving the transcendental equation: $\sinh(9q) - 10q = 0$.

```
> restart;
> q:=fsolve(sinh(9*q)-10*q=0,q,q=0..1);
```
$$q := 0.08927066978$$

The catenary curve is $y(x) = (1/q)(\cosh(qx)-1)$.

```
> To:=40/q:
> y:=x->(1/q)*(cosh(q*x)-1):
> plot([x,y(x),x=-9..9],x=-10..10,y=-5..5,
  labels=[`x`,`y(x)`],labelfont=[TIMES,BOLD,12],
  color=black);
```

Cables

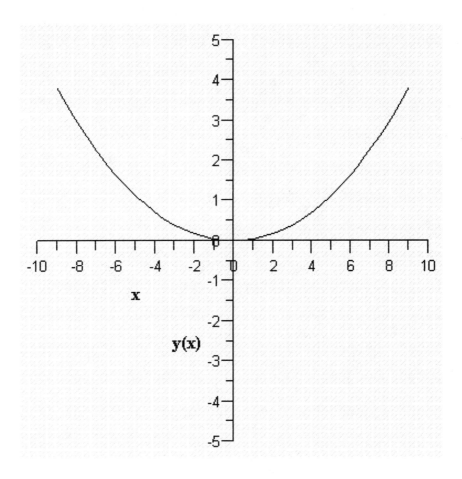

The tension in the cable is given by:

```
> T:=x->To*cosh(q*x):
> plot([x,T(x),x=-9..9],x=-10..10,y=400..700,
  labels=[`x`,`T(x)`],labelfont=[TIMES,BOLD,12],
  color=black);
```

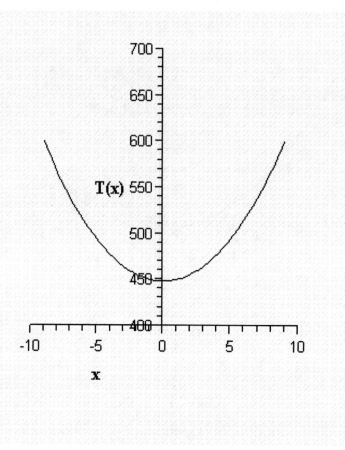

The maximum tension is:

> **maxT :=T(9);**

$$maxT := 600.6427003$$

9 Friction

Coulomb friction or dry friction is an important concept. The coefficient of static friction is the ratio of the maximum friction force to the normal force between the two contacting surfaces. As in any equilibrium problem, the solution will involve a system of simultaneous equations and *Maple* can be used to solve the resulting system.

Computational Solution — Sample Problem 9.2

The numerical work for Sample Problem 9.2 can easily be done using *Maple*. The two linear equations for *P* and *N* will be written in matrix notation and solved twice, once for a positive friction force and once for a negative friction force.

The coefficient matrices will be different for the maximum and minimum cases.

```
> restart;
> with(linalg):
> Cmin:=matrix(2,2,[cos(25*Pi/180),.2,-sin(25*Pi/180),1]);
> Cmax:=matrix(2,2,[cos(25*Pi/180),-.2,-sin(25*Pi/180),1]);
```

$$Cmin := \begin{bmatrix} \cos\left(\frac{5}{36}\pi\right) & 0.2 \\ -\sin\left(\frac{5}{36}\pi\right) & 1 \end{bmatrix}$$

$$Cmax := \begin{bmatrix} \cos\left(\frac{5}{36}\pi\right) & -.2 \\ -\sin\left(\frac{5}{36}\pi\right) & 1 \end{bmatrix}$$

The column matrix for the right side of the equation is the same for both cases.

```
> RS:=matrix(2,1,[400*sin(25*Pi/180),400*cos(25*Pi/180)]);
```

$$RS := \begin{bmatrix} 400\sin\left(\frac{5}{36}\pi\right) \\ 400\cos\left(\frac{5}{36}\pi\right) \end{bmatrix}$$

```
> [Pmin,Nmin]=transpose(evalf(multiply(inverse(Cmin),RS)));
```

$$([Pmin, Nmin]) = \begin{bmatrix} 97.43602989 & 403.7013604 \end{bmatrix}$$

```
> [Pmax,Nmax]=transpose(evalf(multiply(inverse(Cmax),RS)));
```
$$([Pmax, Nmax]) = \begin{bmatrix} 293.9359833 & 486.7458292 \end{bmatrix}$$

The range of P for no slip is $97.4 < P < 293.9$ pounds.

In Sample Problem 9.2, the force required to push the block up the incline may be computed as a function of the angle of incline for example (0 to 45 degrees) and the coefficient of friction (0 to 0.7) and the results displayed on an appropriate surface plot.

Computational Solution — Sample Problem 9.2 — Solution as a Function of the Coefficient of Friction and the Inclination Angle of the Plane

```
> restart;
> with(linalg):
> C_max:=(theta,mu)->matrix(2,2,[cos(theta),-mu,
  -sin(theta),1]):
> C_max_inverse:=(theta,mu)->inverse(C_max(theta,mu)):
> RS:=(theta,mu)->
  matrix(2,1,[400*sin(theta),400*cos(theta)]):
> maxforces:=(theta,mu)->
  multiply(C_max_inverse(theta,mu),RS(theta,mu)):
> p:=(theta,mu)->evalf(maxforces(theta,mu)[1,1]):
> plot3d(p(theta*Pi/180,mu),theta=0..45,mu=0..7/10,
  view=0..2500,labels=[`theta(degrees)`,`mu`,`p`],
  labelfont=[TIMES,BOLD,12],shading=zgrayscale,
  axes=boxed,orientation=[300,60]);
```

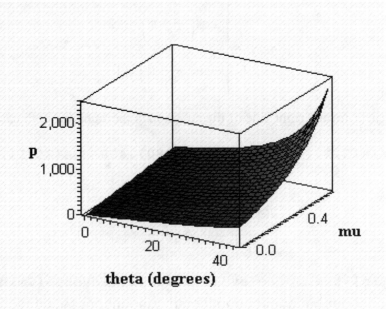

The maximum P (in pounds) occurs when θ is 45 degrees and the coefficient of friction is 0.7.

> **pmax:=p(45*Pi/180,.7);**

$$pmax := 2266.666668$$

If the incline were at an angle of 45 degrees and the coefficient of friction were 0.7, the horizontal force required to push the 400 pound box up the incline would be 2267 pounds. Analyses and surface plots of this type are used by engineers during the design stage.

Computational Solution — Sample Problem 9.3

Sample Problem 9.3 is solved using *Maple* to determine the minimum force and the angle of application of the force for a specific incline angle.

Determine the minimum force necessary to initiate movement of the block up the incline.

```
> restart;
> W:=200: alpha:=25*Pi/180: u:=.2:
> P:=theta->(W*(sin(alpha)+u*cos(alpha)))/
  (cos(theta-alpha)+u*sin(theta-alpha)):
> plot([theta,P(theta),theta=0..2],x=0..2,y=100..200,
  labels=[`theta(radians)`,`P(theta)`],
  labelfont=[TIMES,BOLD,12],color=black);
```

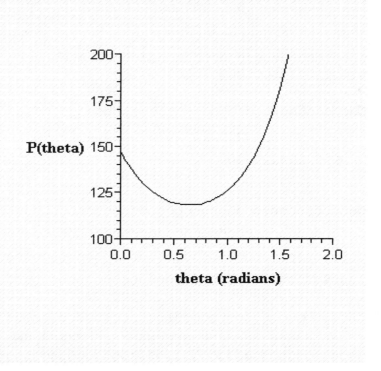

```
> dP:=theta->diff(P(theta),theta):
> plot([theta,dP(theta),theta=0..2],x=0..2,
  y=-200..400,labels=[`theta(radians)`,`dP(theta)`],
  labelfont=[TIMES,BOLD,12],color=black);
```

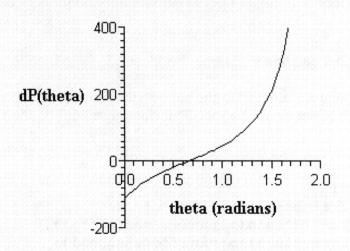

The minimun force is defined by the location of zero slope on the $P(\theta)$ versus θ curve. The location of zero slope in radians is:

> **zero_slope:=fsolve(dP(theta)=0,theta);**
$$zero\_slope := 0.6337278728$$

> **evalf(0.633728*180/Pi);**
$$36.30993975$$

> **maxP:=evalf(P(zero_slope));**
$$maxP := 118.4305762$$

Computational Solution — Sample Problem 9.4

Sample Problem 9.4 can be solved using *Maple* as a vector calculator for any numerical case. Consider the case when $a = 25$ degrees and $b = 40$ degrees and the mass is 40 kg. Determine the minimum coefficient of friction to prevent slipping.

Note the fact that the magnitude of a vector is equal to the square root of the scalar product of the vector with itself is used in the following sequence.

Wedges

```
> restart;
> with(linalg):
> alpha:=25*Pi/180:   beta:=40*Pi/180:   g:=9.81:
> m:=vector([cos(alpha),0,-sin(alpha)]):
> q:=vector([0,cos(beta),-sin(beta)]):
> n:=evalf(evalm(crossprod(m,q)/
  sqrt(dotprod(crossprod(m,q),crossprod(m,q)))));
```
$$n := \begin{bmatrix} 0.3363944699 & 0.6053266994 & 0.7214002686 \end{bmatrix}$$

```
> W:=vector([0,0,-40*g]):
> Wn:=evalm((dotprod(W,n))*n);
```
$$Wn := \begin{bmatrix} -95.22569391 & -171.3543478 & -204.2121596 \end{bmatrix}$$

```
> N:=evalm(-Wn):
> Wt:=evalm(W-Wn):
> f:=evalm(-Wt);
```
$$f := \begin{bmatrix} -95.22569391 & -171.3543478 & 188.1878404 \end{bmatrix}$$

The coefficient of static friction is:

```
> u:=sqrt(dotprod(f,f))/sqrt(dotprod(N,N));
```
$$u := 0.9599640738$$

These vector calculations are easily done using any form of a vector calculator.

Wedges

The general solution of the four equations for the wedge system shown in Figure 9.10 can be solved symbolically using *Maple*. The coefficient matrix is entered and the inverse of this matrix is determined using the *inverse* function. The solution is then determined symbolically using the methods of linear algebra.

Computational Solution — Figure 9.10 — Symbolic Evaluation of Wedge System

```
> restart;
> with(linalg):
> coefficient_matrix:=matrix(4,4,[0,-1,mu2,0,0,-
  mu1,1,0,1,0,-mu2,-(sin(theta)+mu3*cos(theta)),0,0,-
  1,cos(theta)-mu3*sin(theta)]);
```

$$coefficient\_matrix := \begin{bmatrix} 0 & -1 & \mu 2 & 0 \\ 0 & -\mu 1 & 1 & 0 \\ 1 & 0 & -\mu 2 & -\sin(\theta) - \mu 3 \cos(\theta) \\ 0 & 0 & -1 & \cos(\theta) - \mu 3 \sin(\theta) \end{bmatrix}$$

> `constant_matrix:=matrix(4,1,[0,W,0,0]);`

$$constant\_matrix := \begin{bmatrix} 0 \\ W \\ 0 \\ 0 \end{bmatrix}$$

> `multiply(inverse(coefficient_matrix),constant_matrix);`

$$\begin{bmatrix} -\dfrac{(-\mu 2\cos(\theta) + \mu 2\,\mu 3\sin(\theta) - \sin(\theta) - \mu 3\cos(\theta))\,W}{\cos(\theta) - \mu 3\sin(\theta) - \mu 1\,\mu 2\cos(\theta) + \mu 1\,\mu 2\,\mu 3\sin(\theta)} \\ -\dfrac{\mu 2\,W}{-1 + \mu 1\,\mu 2} \\ -\dfrac{W}{-1 + \mu 1\,\mu 2} \\ \dfrac{W}{\cos(\theta) - \mu 3\sin(\theta) - \mu 1\,\mu 2\cos(\theta) + \mu 1\,\mu 2\,\mu 3\sin(\theta)} \end{bmatrix}$$

Consider the case when the coefficient of static friction is the same for all surfaces and is equal to 0.2. The applied force P is linearly related to the weight W. Therefore, the weight will be treated as a unit weight times 100% and the resulting normal forces and the applied force will be percentages of W. The four linear equations will be solved using matrix algebra.

Computational Solution — Figure 9.10 — Dependency of the Applied Force P on the Wedge Angle θ

> `restart;`
> `with(linalg):`
> `mu:=0.2:`
> `coefficient_matrix:=theta->matrix(4,4,[0,-1,mu,0,0,-mu,1,0,1,0,-mu,-(sin(theta)+mu*cos(theta)),0,0,-1,cos(theta)-mu*sin(theta)]):`
> `constant_matrix:=theta->matrix(4,1,[0,100,0,0]):`
> `F:=theta->multiply(inverse(coefficient_matrix(theta)),constant_matrix(theta)):`
> `P:=theta->F(theta)[1,1]:`

The first component of the vector F is the applied force P, the second is N_1, the third is N_2, and the last is N_3. Plotting the value of P versus wedge angle yields the following.

> `plot([theta,P(theta*Pi/180),theta=1..20],x=0..20,`
  `y=40..100,labels=[`theta (degrees)`,`P(theta)`],`
  `labelfont=[TIMES,BOLD,12],color=black);`

Wedges

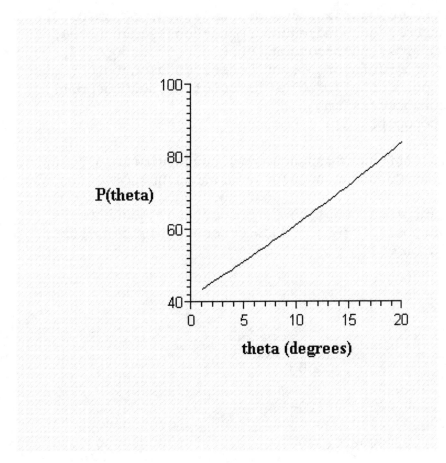

The following are *P* values as percentages of *W*.

```
> P_1deg:=evalf(P(1*Pi/180));
> P_20deg:=evalf(P(20*Pi/180));
```

$$P\_1deg := 43.56425655$$

$$P\_20deg := 84.19239525$$

In this case, the wedge angle was varied from one degree to 20 degrees and the required wedge force went from 43.6% of the weight to 84.2% of the weight. The greater wedge angle would lift the weight higher than the smaller wedge angle.

Computational Solution — Figure 9.10 — Dependency of the Applied Force *P* on the Coefficient of Friction μ

A specific wedge angle is chosen for this example.

```
> restart;
> with(linalg):
> theta:=10*Pi/180:
```

```
> coefficient_matrix:=mu->matrix(4,4,[0,-1,mu,0,0,-
  mu,1,0,1,0,-mu,-(sin(theta)+mu*cos(theta)),0,0,-
  1,cos(theta)-mu*sin(theta)]):
> constant_matrix:=mu->matrix(4,1,[0,100,0,0]):
> F:=mu->multiply(inverse(coefficient_matrix(mu)),
  constant_matrix(mu)):
> P:=mu->F(mu)[1,1]:
```

The coefficient of friction for all surfaces will vary from 0.0 to 0.6. The following is a plot of the force P as a percentage of W versus the coefficient of friction.

```
> plot([mu,P(mu),mu=0..6/10],x=0..6/10,y=0..300,
  labels=[`mu`,`P(mu)`],labelfont=[TIMES,BOLD,12],
  color=black);
```

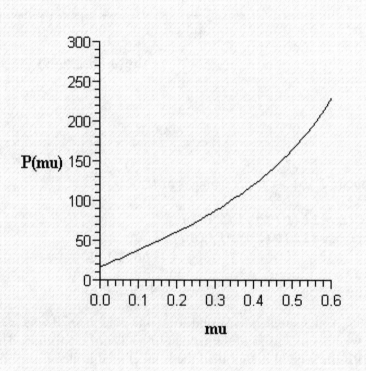

```
> P_0mu:=evalf(P(0));
> P_point6mu:=evalf(P(.6));
```

$$P\_0mu := 17.63269807$$

$$P\_point6mu := 229.4026211$$

It may be easily seen that wedges are very sensitive to the coefficient of friction between the contacting surfaces. For smooth surfaces, the force P required to raise the weight is 17.6% of the weight while with a coefficient of friction of 0.6, the force P is 229.4% of the weight. Although friction appears to be detrimental to the use of the wedge, some friction is necessary to hold the wedge in place.

Belt Friction

We can gain a better understanding of the belt friction discussed in Section 9.5 in the text by plotting Eq. (9.25) as we vary different parameters.

Ratio of Tensions versus the Coefficient of Friction

```
> restart;
> beta:=1/2*Pi:
```

The ratio of T_2 to T_1 will be represented by the letter R.

```
> R:=u->exp(u*beta):
> plot([u,R(u),u=0..1],x=0..1.2,y=0..6,labels=[`u`,`R(u)`],
  labelfont=[TIMES,BOLD,12],color=black);
```

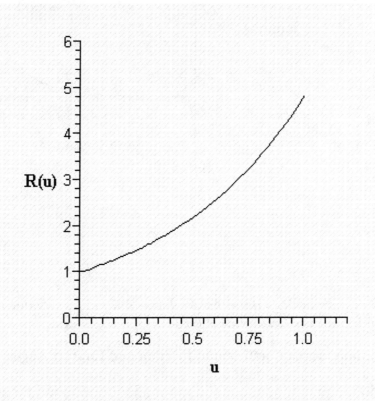

It is also of interest to investigate the effect of increasing the contact angle. A plot of the ratio of tensions versus contact angle varying from zero to wrapped around the drum two and one half times can easily be generated in *Maple*.

Ratio of Tensions versus the Angle of Contact

```
> restart;
> u:=.4:
> R:=beta->exp(u*beta):
> plot([beta,R(beta),beta=0..5*Pi],x=0..16,y=-500..1000,
    labels=[`beta(radians)`,`R(beta)`],
    labelfont=[TIMES,BOLD,12],color=black);
```

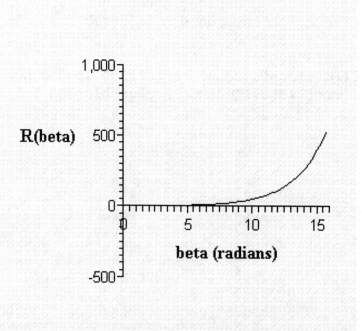

A surface plot can be made to show the full dependency of belt friction on both the coefficient of friction and the contact angle β.

Ratio of Tensions versus Coefficient of Friction and Contact Angle

```
> restart;
> R:=(u,beta)->exp(u*beta):
> plot3d(R(u,beta),u=0..1/2,beta=0..5*Pi/2,
    labels=[`u`,`beta(radians)`,`R`],
    labelfont=[TIMES,BOLD,12],shading=zgrayscale,
    axes=boxed,orientation=[10,60]);
```

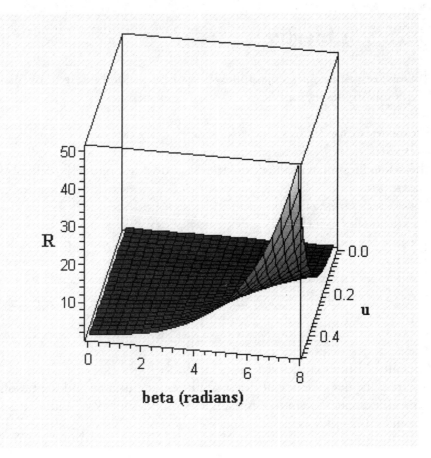

The maximum R occurs when $\mu = 0.5$ and $\beta = 430$ degrees.

```
> Rmax:=evalf(R(.5,5*Pi/2));
```
$$Rmax := 50.75401956$$

Surface plots are useful to visually see the dependency of an analysis on the variables involved. The plot shown was done for coefficients of friction from 0 to 0.5 and for values of β from 0 to 430 degrees.

10 Moments of Inertia

Many of the integrals arising in the computation of the second moment of the area can be evaluated symbolically using *Maple*.

Computational Window 10.1

Consider the second moment of the circular area shown in Figure 10.5 about its centroidal axis.

Symbolically evaluate the integral.

```
> restart;
> int(int(r^3*(sin(theta))^2,r=0..R),theta=0..2*Pi);
```

$$\frac{1}{4} R^4 \pi$$

Principal Second Moments of Area

It is useful to plot the dependency of the second moment of area and the product moment of area on the coordinate system rotation angle. Computational Window 10.2 is a plot of Eqs. (10.33) and (10.34).

Computational Window 10.2 — Second Moment of Area

```
> restart;
> with(plots):
> Ixx:=10: Iyy:=5: Ixy:=0:
> pIxx:=beta->.5*(Ixx+Iyy)+.5*(Ixx-
  Iyy)*cos(2*beta)+Ixy*sin(2*beta):
> pIxy:=beta->-.5*(Ixx-Iyy)*sin(2*beta)+Ixy*cos(2*beta):
```

The following is a plot of $I_{x'x'}$ and $I_{x'y'}$ versus the angle β.

```
> plot1:=plot([beta,pIxx(beta),beta=0..2*Pi],x=0..7,
  y=-10..20,color=black):
> plot2:=plot([beta,pIxy(beta),beta=0..2*Pi],x=0..7,
  y=-10..20,color=black):
> textplot1:=textplot([3.5,12,"pIxx(beta)"]):
> textplot2:=textplot([3.5,3,"pIxy(beta)"]):
> display(plot1,plot2,textplot1,textplot2,labels=[`beta
  (radians)`,``],labelfont=[TIMES,BOLD,12]);
```

Principal Second Moments of Area

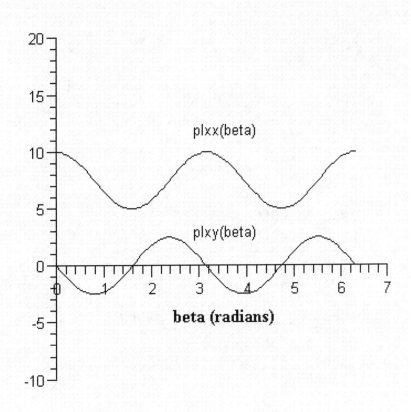

Computational Solution — Sample Problem 10.5

```
> restart;
> with(plots):
> Ixx:=136: Iyy:=64: Ixy:=-48:
> pIxx:=beta->.5*(Ixx+Iyy)+.5*(Ixx-
  Iyy)*cos(2*beta)+Ixy*sin(2*beta):
> pIxy:=beta->-.5*(Ixx-Iyy)*sin(2*beta)+Ixy*cos(2*beta):
```

The following is a plot of $I_{x'x'}$ and $I_{x'y'}$ versus the angle β.

```
> plot1:=plot([beta,pIxx(beta),beta=0..2*Pi],x=0..7,
  y=-100..200,color=black):
> plot2:=plot([beta,pIxy(beta),beta=0..2*Pi],x=0..7,
  y=-100..200,color=black):
> textplot1:=textplot([4.25,125,"pIxx(beta)"]):
> textplot2:=textplot([3.1,50,"pIxy(beta)"]):
> display(plot1,plot2,textplot1,textplot2,
  labels=[`beta (radians)`,``],labelfont=[TIMES,BOLD,12]);
```

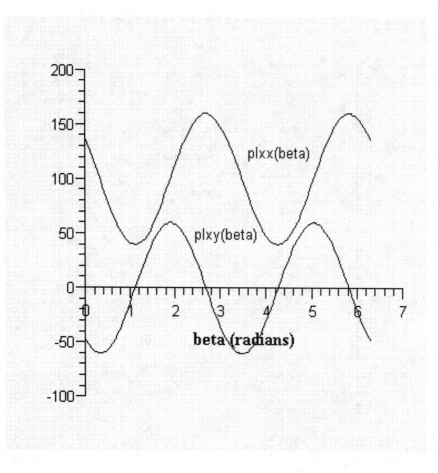

Eigenvalue Problem

The eigenvalue problem is presented in Section 10.10 in the text. *Maple* can find the eigenvalues and eigenvectors of a square matrix using the *eigenvals(A,vecs)* function where *A* is a square matrix. The eigenvalues are returned when the command is evaluated. The eignevectors are stored columnwise in the matrix *vecs*. The matrix *vecs* is not automatically returned. It can be viewed using the command *print(vecs)*. The first column of the *vecs* matrix corresponds to the first eigenvalue. This correspondence holds for the remainder of the eigenvalues and eigenvectors.

Computational Solution — Sample Problem 10.7 — Eigenvalue Solution

We will use the numerical values from Sample Problem 10.5 to determine the principal axes.

The following is the second moment of inertia matrix. Note that the character *I* cannot be used as a variable (it is reserved by *Maple* for the imaginary number defined as the square root of negative one).

```
> restart;
> with(linalg):
> I2:=matrix(2,2,[136,-48,-48,64]):
```

```
> lambda:=evalf(Eigenvals(I2,vecs));
```
$$\lambda := \begin{bmatrix} 40.0000000 & 160.0000000 \end{bmatrix}$$

```
> print(vecs);
```
$$\begin{bmatrix} -.4472135956 & -.8944271912 \\ -.8944271912 & 0.4472135956 \end{bmatrix}$$

The first eigenvector, which corresponds to the eigenvalue 160 is:

```
> eigenvector1:=vector([vecs[2,1],vecs[2,2]]);
```
$$eigenvector1 := \begin{bmatrix} -.8944271912 & 0.4472135956 \end{bmatrix}$$

The second eigenvector, which corresponds to the eigenvalue 40 is:

```
> eigenvector2:=vector([vecs[1,1],vecs[1,2]]);
```
$$eigenvector2 := \begin{bmatrix} -.4472135956 & -.8944271912 \end{bmatrix}$$

This is 180 degrees from the answer obtained in Sample Problem 10.5 but the axis is still the same. Note that if a square matrix is symmetric and has distinct eigenvalues (as is the case in Sample Problem 10.5), the eigenvectors of the matrix are orthogonal and the dot product of orthogonal vectors is zero.

```
> dotprod(eigenvector1,eigenvector2);
```
$$0$$

Note that in *Maple* elements of a vector are referenced using the syntax *vector[element]* where the first element has index *[1]*. Elements of matrices are referenced using the syntax *matrix[row][column]* where the upper left entry of the matrix has index *[1][1]*.

11 Virtual Work

The work done by a particle is equal to the integral of the dot product of the force acting on the particle with the differential change in the position vector. *Maple* can be used to symbolically or numerically evaluate some of the integrals that arise.

Computational Solution — Sample Problem 11.3

The integrals in Sample Problem 11.3 must be evaluated numerically because no exact forms exist. Referring to the special functions found in a table of integrals, the integrals in this sample problem will involve Fresnel sine and cosine integrals.

```
> restart;
> with(plots):
> theta:=s->1+s^2:
> x:=s->int(cos(theta(u)),u=0..s):
> y:=s->int(sin(theta(u)),u=0..s):
> plot1:=plot([s,x(s),s=0..4],x=0..5,y=-1..2,color=black):
> plot2:=plot([s,y(s),s=0..4],x=0..5,y=-1..2,color=black):
> textplot1:=textplot([2.25,1,"x(s)"],
    labelfont=[TIMES,BOLD,12]):
> textplot2:=textplot([1.5,-.5,"y(s)"],
    labelfont=[TIMES,BOLD,12]):
> display(plot1,plot2,textplot1,textplot2,labels=[`s`,``],
    labelfont=[TIMES,BOLD,12]);
```

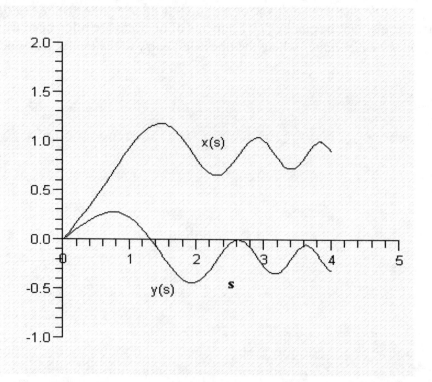

Computational Solution — Sample Problem 11.3

```
> plot([x(s),y(s),s=0..4],x=-6/10..4/10,y=0..3/2,
  labels=[`x(s)`,`y(s)`],labelfont=[TIMES,BOLD,12],
  color=black);
```

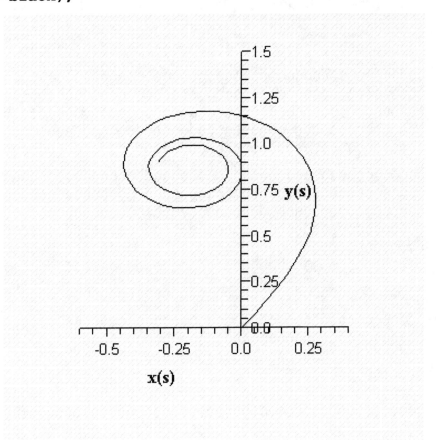

The work done by the constant force of 50 N acting in the *x*-direction is determined by the following analysis.

```
> f:=s->50:
> beta:=s->0:
> U:=s->int(f(u)*cos(theta(u)-beta(u)),u=0..s):
> plot([s,U(s),s=0..4],x=0..5,y=-40..20,
  labels=[`s`,`U(s)`],labelfont=[TIMES,BOLD,12],
  color=black);
```

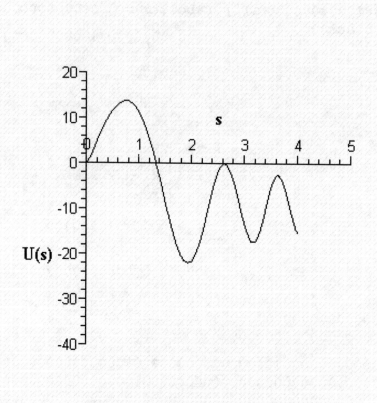

The total work done in N · m is given by the value of U at the end of the path (s = 4 m).

> **Utot:=evalf(U(4));**

$$Utot := -15.37515828$$

Notice that the force does both positive and negative work as it moves along the path and at 4 meters the total work done is negative.

Similar integrals arise in Sample Problem 11.4. The numerical solutions for these integrals are shown.

Computational Solution — Sample Problem 11.4 — Slide Problem

Consider a slide that has an initial angle of 60 degrees downward and a final angle of zero degrees. The path of the slide is parabolic.

```
> restart;
> n:=2:
> theta:=s->(-60*Pi/180)*(1-s^n):
> x:=s->int(cos(theta(u)),u=0..s):
> y:=s->int(sin(theta(u)),u=0..s):
```

Computational Solution — Sample Problem 11.4 — Slide Problem

```
> plot([x(s),y(s),s=0..1],x=0..8/10,y=-.8..0,
  labels=[`x(s)`,`y(s)`],labelfont=[TIMES,BOLD,12],
  color=black);
```

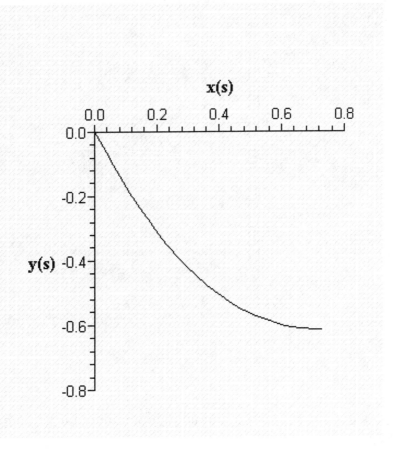

The total work done for a slide of length 1 is:

```
> m:=20: u:=.3: W:=m*9.81:
> U:=s->int(-W*sin(theta(v))-u*W*cos(theta(v)),v=0..s):
> plot([s,U(s),s=0..1],x=0..1,y=0..100,labels=[`s`,`U(s)`],
  labelfont=[TIMES,BOLD,12],color=black);
```

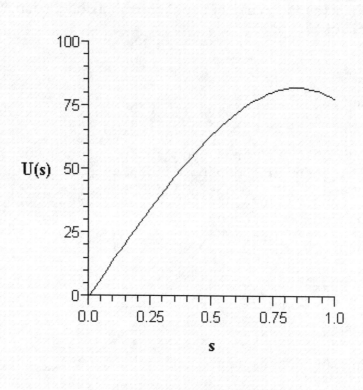

The total work done for the normalized slide in Nm is given by the value of U at the end of the path ($s = 1$ m).

```
> Utot:=U(1);
```

$$Utot := 77.74079875$$

To determine the total work done for the original slide, multiply $U(1)$ by the length of the slide. This length may be determined from the height of the slide which for this case is approximately $0.6s$. Note that when dynamics is considered, the velocity of the 20 kg child at the bottom of the slide could be determined.

Computational Solution — Sample Problem 11.4 — Slide Problem (Variation)

Another variation of this problem is to examine the effect of the coefficient of friction. As the coefficient of friction increases, more negative work will be done and the child will slow down. When the total work done is zero, the child will stop.

```
> restart;
> n:=2: m:=20: u:=.9: W:=m*9.81:
> theta:=s->(-60*Pi/180)*(1-s^n):
```

```
> U:=s->int(-W*sin(theta(v))-u*W*cos(theta(v)),v=0..s):
> plot([s,U(s),s=0..1],x=0..1,y=-20..40,
  labels=[`s`,`U(s)`],labelfont=[TIMES,BOLD,12],
  color=black);
```

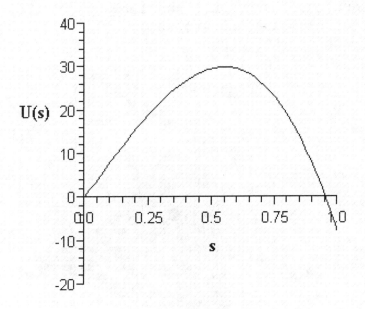

```
> fsolve(U(s)=0,s,s=.5..1);
```
$$0.9527530789$$

```
> U(0.953);
```
$$-0.03873753961$$

The total work done is zero after sliding down 95.3% of the length of the slide and the child would stop at this point. It will be shown in *Dynamics* that the total work done is related to the change in the square of the velocity. Therefore, since the child started at the top of the slide with zero velocity, the velocity will be a maximum after the child has slid a little over 50% of the length of the slide and zero when the child has slid 95.3% of the length of the slide. Solutions of this type of problem are only feasible with use of computational software.

Computational Solution — Sample Problem 11.7

A surface plot for the function in Sample Problem 11.7 shows the general shape of the function.

```
> restart;
> M:=(x,y)->x^2+3*x*y+y^2:
> plot3d(M(x,y),x=-2..2,y=-2..2,labels=[`x`,`y`,`M(x,y)`],
  labelfont=[TIMES,BOLD,12],shading=zgrayscale,
  axes=boxed,orientation=[10,60]);
```

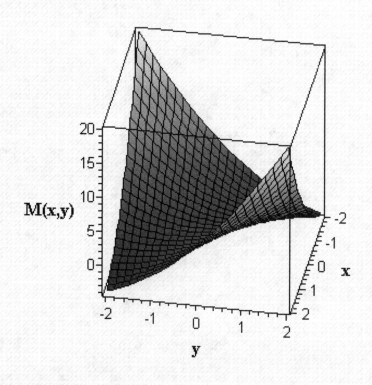

Summary

This supplement shows how to use *Maple* to solve problems in Statics. Most of these problems may be solved by hand but *Maple* reduces the labor of numerical calculations. The nonlinear problems cannot be solved without the use of some computational aid and are examples of the increased capabilities of analysis when computational software is used. This supplement is not meant to replace the manual for *Maple* or to show all the features of the software. Note that *Maple* is a tool for computation and is not a Statics software package. All problems must be modeled and the equations of equilibrium written before the software is used.

Index

A

Algebraic equations, solving with Maple, 6–7
Analysis of structures, *see* Structures
Angle of contact, belt friction, 110–111
Applied forces, friction
    in wedges and, 106–107, 107–109
array function, 3

B

Belt friction, 109–111
    angle of contact and, 110–111
    coefficient of static friction and, 109–110, 110–111
    contact angle and, 110–111
    ratio of tensions, 109–111

C

Cables, 38–39, 39–40, 40–43, 72–73, 96–100
    catenary, 98–100
    changing length of, 96
    deformable, 38–39
    equilibrium and, 72–73
    fsolve function, 96–97
    internal forces and, 96–100
    nonlinear algebraic equations of, 38–39, 39–40
    particle equilibrium and, 38–39, 39–40, 40–43
    statically indeterminate problems, 40–43
    tension, numerical solution of, 72–73
    vertical deformation of, 39–40
Cartesian coordinate system, plotting
    two-dimensional, 7–9
Center of gravity, distributed forces of, 50–55
Centroids, distributed forces of, 50–55
Coefficient matrix, 105–106
Coefficient of static friction, 101, 102–103, 104–105,
    106, 107–109, 109–110, 110–111, 120–121
    applied force, dependency of on, 106, 107–109
    belt friction and, 109–110, 110–111
    contact angle and, ratio of tensions versus, 110–111
    defined, 101
    inclination angle and, 102–103
    ratio of tensions versus, 109–110
    wedges and, 106, 107–109
    virtual work and, 120–121
Computational generation of equilibrium
    equations, 65–66
Contact angle, belt friction, 110–111
coords option, 8
Cross product, 26–30
    crossprod function, 26–27
    vector analysis and, 26–30
crossprod function, 26–27

D

Deformation analysis using nonlinear algebraic
    equations, 36–43
diff function, 3

Dirac delta function, internal forces and, 92
Dirac(t) function, 92
Direct vector solution, 33–34
    equivalent force systems, 44–45
    particle equilibrium, 33–34
Discontinuity functions, 92–96
    Dirac delta function, 92
    Heaviside step function, 92–93
    internal forces and, 92–96
    shear and moment diagrams, 93–96
display command, 54–55
Distributed forces, 50–55
    centroids, 50–55
    center of gravity, 50–55
    numerical integration, 50, 51–52
    symbolic integration, 50–51
    scatter plots, three dimensional, 53–55
Document mode, 1
Dot product, 16, 23–26
    dotprod function, 16, 23–24
    evalf function, 23–24
    vector analysis and, 23–26
    vector calculator, use in a, 16
dotprod function, 16, 23–24

E

eigenvals(A,vecs) function, 114–115
Eigenvalue problem, 114–115
Equal sign (=), 7
Equilibrium, 31–43, 56–73. *See also* Particle equilibrium
    cable tension, numerical solution of, 72–73
    equations, 36–37, 59–71
    evalm function, 60
    fsolve function, 56–58, 61
    particle, 31–43
    rigid bodies, 56–73
    root determination using *fsolve* function, 56–58
    solve function, 61, 62
    summation of forces for, 60
    summation of moments for, 60–61
    symbolic (processor) generation of equations
        for, 59, 60–61, 62–65, 69–71
Equilibrium equations, 36–37, 59–71, 82–83
    analysis of structures using, 82–83
    computational generation of, 65–66
    fsolve function, 61
    graphing, 67–68
    matrix function, 61–62, 69
    numerical solution of, 71
    particle equilibrium, 36–37
    rigid body equilibrium, 60–71
    simplify function, 63
    solution of, 61–62
    solve function, 61, 62
    symbolic (processor) generation of, 59,
        60–61, 62–65, 69–71

Equivalent force systems, 44–49
   direct vector solution for, 44–45
   moment equations, 45–47
   moment of a force, 44
   wrench, calculations for, 48–49
evalf function, 3–5, 13–14, 23–24
   scalar (dot) product, calculation using, 23–24
   symbolic calculations using Maple, 5
   vector analysis using, 13–14
   working with in Maple, 3–5
evalm function, 60
expand function, 6

F

for loop, 3, 51–52
   distributed forces, determination of using, 51–52
   numerical integration using, 51–52
   od term, 51
   working with in Maple, 3
Forces, 44–49, 50–55, 60, 84–100, 103–104, 106–107, 107–109. *See also* Friction; Tension
   angle of application of, 103–104
   applied, 106–107, 107–109
   distributed, 50–55
   equivalent systems, 44–49
   evalm function, 60
   friction and, 103–104, 106–107, 107–109
   internal, 84–100
   minimum, 103–104
   moment of, 44
   structural members, in, 84–100
   summation of for equilibrium, 60
Friction, 101–111
   angle of application of force, 103–104
   belt, 109–111
   coefficient of static, 101, 102–103, 104–105, 106, 107–109, 109–110, 110–111
   inclination angle, 102–103
   matrix notation, 101–102
   minimum force, 103–104
   ratio of tensions, 109–111
   slipping, prevention of, 104
   wedges, 105–109
fsolve function, 5, 6–7, 13–14, 34–36, 40, 56–58, 61, 78–79, 96–97
   algebraic equations, solving with, 6–7
   analysis of structures using, 78–79
   equilibrium equations and, 61
   equilibrium of rigid bodies using, 56–58, 61
   internal forces in cables, determination of, 96–97
   nonlinear algebraic equations, solving with, 34–36, 40
   root determination using, 56–58
   statically indeterminate problems, solving with, 40
   symbolic calculations using Maple, 5
   vector analysis using, 13–14

Functions, 2–5, 6–7
   array, 3
   diff, 3
   evalf, 3–5
   for loop, 3
   fsolve, 5, 6–7
   subs, 3–4
   working with in Maple, 2–5

G

Graphs, 7, 67–68, 78, 80–81
   analysis of structures using, 78, 80–81
   equilibrium equations solutions using, 67–68
   internal forces in trusses, 78
   Maple, use of in, 7
   motor torque, 80–81

H

Heaviside step function, internal forces and, 92–93
Heaviside(t) function, 92, 93–95
Help drop-down menu, 1

I

Inclination angle, coefficient of static friction and, 102–103
Insert drop-down menu, 1
Int or *int* function, 5, 50, 51
   capitalization of in Maple, 51
   distributed forces, determination of using, 50, 51
   numerical (*Int*) integration using, 50, 51
   symbolic (*int*) calculations, 5
Integral evaluation ($\int$), 1
Integration, 50–52
   distributed forces and, 50–52
   double integrals, 50
   for loops, 51
   int or *Int* operators, 50, 51
   numerical, 50, 51–52
   symbolic, 50–51
Internal forces, 78, 84–100
   cables and, 96–100
   discontinuity functions for, 92–96
   graph of in trusses for structural analysis, 78
   linear loading, numerical solution for, 84–86
   maximum moment, determination of, 88–89
   shear and moment diagrams for, 84, 90–92, 93–96
   sinusoidal loading, numerical solution of, 86–88
   structural members, in, 84–100
inverse function, 105–106

L

linalg package, 15, 16
Linear equations, solution of simultaneous, 20–22
Linear loading, numerical solution for internal forces from, 84–86
Loading function, 22

Index

M

Maple, 1–12, 15–20, 22–23
    algebraic equations, solving with, 6–7
    application to a Statics problem, 9–12
    computational software, use of, 1–12
    Document mode, 1
    equal sign (=), 7
    functions, working with, 2–5
    graphs, 7
    Help drop-down menu, 1
    Insert drop-down menu, 1
    matrix calculations, using for, 22–23
    numerical calculations using, 2
    plots, 7–9
    symbolic calculations using, 5–6
    2-D Input function, 1
    vector calculator, as a, 15–20
    View drop-down menu, 1
    Worksheet mode, 1
Matrix calculations, using Maple for, 22–23
matrix function, 15, 61–62, 69
    equilibrium equations using, 61–62, 69
    vector calculator use of, 15
Matrix method, analysis of structures using, 74–75
Maximum moments, internal forces, 88–89
Method of sections, analysis of structures using, 78
Moment of force, 44
Moments, summation of for equilibrium, 60–61
Moments of inertia, 112–115
    eigenvalue problem, 114–115
    principal second moment of area, 112–114
    print(vecs) function, 114
    second moment of the area, 112
    vecs matrix, 114
Motor torque graph, analysis of structures using, 80–81

N

Nonlinear algebraic equations, 34–43
    cables, computational solution for, 38–39, 39–40
    deformation analysis using, 36–43
    equilibrium equations, 36–37
    fsolve function, 34–36, 40
    particle equilibrium and, 34–43
    procedure for solution of, 34–35
    solution of, 34–43
    spring constants and, 37–38,
    spring deflections and, 36–37
    statically indeterminate problems, 40
    tension and, 36–37
normalize operator, 17
Numerical calculations using Maple, 2
Numerical integration, 50, 51–52
    distributed forces and, 50, 51–52
    for loops, 51–52
    Int operator, 50, 51
Numerical solution of equilibrium equations, 71
numpoints option, 7

O

Operators, *see* Functions

P

Particle equilibrium, 31–43
    cables, computational solution for, 37–38, 39–40
    direct vector solution for, 33–34
    equilibrium equations, 36–37
    fsolve function, 34–36
    nonlinear algebraic equations, solution of, 34–43
    parametric solutions for, 3134
    soft springs, computational solution for, 37–38
Plots, 7–9, 53–55, 122
    clearing memory for, 9
    coords option, 8
    creation of in Maple, 7–9
    display command, 54–55
    numpoints option, 7
    plot function, 7–8
    pointplot3d function, 53
    polar, 8–9
    scatter, three-dimensional, 53–55
    surface, 122
    textplot command, 54–55
    two-dimensional Cartesian coordinate system, 7–9
    with(plots) command, 54
pointplot3d function, 53
Polar plots, 8–9
Principal second moment of area, 112–114
print(vecs) function, 114

R

Ratio of tensions, 109–111
    belt friction and, 109–111
    coefficient of friction versus, 109–110
    contact angle and coefficient of friction, versus, 110–111
    contact angle versus, 110
Rigid bodies, 44–49, 56–73
    equilibrium of, 56–73
    equivalent force systems, 44–49
Roots, determination using *fsolve* function, 56–58

S

Scalar (dot) product, 23–26
Scatter plots, 53–55
    distributed forces and, 53–55
    display command, 54–55
    pointplot3d function, 53
    textplot command, 54–55
    three-dimensional, 53–55
    with(plots) command, 54
Second moment of area, 112
Shear and moment diagrams, 84, 90–92, 93–96
    discontinuity functions, 92–96
    internal forces, 84, 90–92, 93–96
simplify function, equilibrium equations, 63

Sinusoidal loading, numerical solution for internal forces from, 86–88
Slipping, prevention of, 104
solve function, equilibrium equations, 61, 62
Springs, 36–37, 37–38
    constants, changing values of, 37–38
    deflections, nonlinear algebraic equations and, 36–37
    particle equilibrium and, 36–37, 37–38
    soft, nonlinear algebraic equations for, 37–38
Static friction, *see* Coefficient of static friction
Statics problem, application to using Maple, 9–12
Structures, 74–83, 84–100
    analysis of, 74–83
    equilibrium equations for, 82–83
    fsolve function, 78–79
    internal forces in members, 84–100
    matrix method, 74–75
    members, internal forces of, 84–100
    method of sections, 78
    motor torque graph for, 80–81
    trusses, 74–78
subs function, 3–4
Surface plot, 122
Symbolic calculations, 5–6
    evalf function, 5
    expand function, 6
    Fsolve function, 5
    Int function, 5
    Maple, using, 5–6
Symbolic integration, 50–51
Symbolic processor, 18, 59, 60–61, 62–65, 69–71
    equilibrium equations, generation of, 59, 60–61, 62–65, 69–71
    vector analysis and, 18

T
Tension, 36–37, 72–73. *See also* Ratio of tensions
    equilibrium in cables, numerical solution of, 72–73
    nonlinear algebraic equations and, 36–37
textplot command, 54–55
Total work, 118, 120
Trusses, 74–78
    graph for internal forces of, 78
    matrix method, 74–75
    method of sections, 78
    structural analysis of, 74–78
2-D Input function, 1

U
Unit vector, creation of, 17

V
vecs matrix, 114
Vector algebra, vector analysis and, 16–17
Vector analysis, 13–30
    cross product, 26–30
    dot product, 23–26
    evalf function, 13–14, 23–24

Fsolve function, 13–14
    linear equations, solution of simultaneous, 20–22
    matrix calculations, using Maple for, 22–23
    procedure for, 13
    scalar (dot) product, 23–26
    vector calculator, Maple as a, 15–20
    vector (cross) product, 26–30
Vector calculator, 15–20
    computational solutions using a, 18–20
    dotprod function, 16
    linalg package, 15, 16
    Maple as a, 15–20
    matrix function, 15
    normalize operator, 17
    symbolic processor, use of, 18
    unit vector, creating with, 17
    vector algebra using a, 16–17
    vector analysis using a, 15–20
    vector function, 15
Vector (cross) product between two vectors, 26–30
vector function, 15
View drop-down menu, 1
Virtual work, 116–122
    coefficient of friction and, 120–121
    determination of, 116–118
    slide problem for, 118–122
    surface plot for, 122
    total, 118, 120

W
Wedge angle, applied force on, 106–107
Wedges, 105–109
    angle, applied force on, 106–107
    applied force, dependency of, 106–107, 107–109
    coefficient matrix, 105–106
    coefficient of static friction, 106, 107–109
    friction and, 105–109
    inverse function, 105–106
    symbolic evaluation of, 105–106
with(plots) command, 54
Work, defined, 116
Worksheet mode, 1
Wrench, calculations for equivalent force systems of, 48–49